Healthcare Automation in Covid 19

Healthcare Automation in Covid 19

Written by Austin Mardon, Sharon He,
Salman Ali, Iqra Chaudhry, Amy Paul
and Catherine Mardon

Typeset and Design by Kiran Rai

GM PRESS

2020

A Golden Meteorite Press Book
Printed in Canada

First Printing: 2020

ISBN 978-1-77369-187-9

Golden Meteorite Press
103 11919 82 St NW
Edmonton, AB T5B 2W3
www.goldenmeteoritepress.com

We acknowledge the support of Canada Service Corps, TakingITGlobal, and the Government of Canada in promotional materials associated with the Project.

Thank you.

Table of Contents

Introduction

COVID-19 is an infectious disease characterized by symptoms such as fever, cough, fatigue, shortness of breath, and loss of smell and taste (HealthLink BC, 2020). Beginning in the city of Wuhan, China, this new coronavirus was considered a mere condition. A spike in cases and its spread from country to country in a short period of time proved us to be wrong. Soon enough, the World Health Organization (WHO) declared COVID-19 a global health emergency.

Quarantines were soon implemented on those who were believed to have contracted the virus. Nations went into full lockdown for an extended period and physical distancing was required in public spaces to prevent the spread of COVID-19. This deadly virus quickly went from being the talk of the town to becoming the talk of the globe. COVID-19's health implications impacted various employment sectors, especially hard hitting medical care systems worldwide. In fact, we are facing a plethora of issues which include inadequate testing, lack of personal protective equipment (PPE), production delays, and shortage of healthcare supplies and professionals.

Testing is important to control the spread of the virus. Although it isn't used to treat patients, it gives us a better idea of the impact that the virus has and to what extent aid would need to be provided. We are testing as many individuals as we possibly can, yet many cases are going unreported. The neverending surge in cases puts healthcare workers and those being tested at risk of contracting or even spreading the virus at a higher rate, making many hesitant to simply go and get tested. Due to national lockdowns and business restrictions, healthcare supplies and PPE are not reaching hospitals and clinics on a timely basis. This is a factor in the large numbers of fatalities that we are seeing as a result of this virus. Medical professionals are not appropriately equipped to serve patients which can put themselves and patients at risk. Furthermore, healthcare professionals including nurses and doctors are becoming medically unfit to perform their roles due to direct exposure with the virus. They are also at the brink of exhaustion, providing endless care to patients aiming to make a recovery. All of these issues are alarming. If we do not find a solution soon, our North American healthcare systems could collapse entirely and we may never be able to ward off the threat of COVID-19. We need to ensure that tests are conducted quickly and safely, that our manufacturing increases, and that we find ways in which healthcare professionals can still provide care even when not in direct contact with patients.

We have to take advantage of living in the 21st century. A century of innovation. We are in need of artificial intelligence (AI) and robotic technologies more than ever. It's the only way that we will be able to overcome this disease.

The unbelievable progress that technology in the health sector has made in the past twenty years ranges from being able to diagnose illnesses quicker to performing surgeries with the help of robots. Work in the field has inspired biotechnology and AI companies across the world to develop new technologies in our fight against COVID-19. These technologies can handle several medical tasks including diagnosis, sample collection, prevention of infection, detection and sanitization of hot-spots, nursing, and manufacturing of healthcare products.

Chapter 3 aims to explore how robotics have been used in the diagnosis of COVID-19. In particular, companies such as Flow Robotics have created pipette robots that can perform over 1,000 tests per day and return results within a maximum of 3 weeks (Flow Robotics, 2020). A technology such as this allows for healthcare workers to conduct as many tests as possible in a safe manner. The development of robotic maps which use CT scan information, is yet another example of automation which has allowed for us to diagnose the disease (Bahl et al., 2020). By taking advantage of the knowledge that we have on the impact that COVID-19 symptoms have on the lungs and nasopharynx, general practitioners can accurately detect whether an individual has the disease and what steps must be taken if tested positive. As part of diagnosis, intelligent thermometry systems have also been under development as discussed in Chapter 4. One such example is Kinsa, a thermometer which can be connected to a mobile application. It provides a database of temperatures and symptoms from individuals all across the country, which allows for government agencies to track susceptible and infected populations over time (Indelicato, 2020). A system such as Kinsa can be used to predict and direct our efforts towards hotspot regions.

To reduce the need for direct patient-doctor contact, robots have been developed to collect coronavirus samples. In Chapter 5, we mention the development of a throat swabbing robot which uses a tip to swab inside of a person's mouth (Crowe, 2020) as well as a venipuncture robot which draws blood samples and analyzes them instantly for COVID-19 antibodies (Balter et al., 2018). Both of these technologies rely on our knowledge of viral and antibody testing and reduce the need for humans to interfere and conduct these procedures. They put both the patient and healthcare workers at less risk of contracting the virus while ensuring that proper, standardized tests are conducted.

AI and similar technologies also have their use in infection prevention (see Chapter 6). UV-C light technology which has been used for many years, can alter and destroy the DNA of SARS-CoV-2, the virus responsible for causing the coronavirus disease, on instant. The THOR UVC robot incorporates this technology along with a mapping system to determine which locations have yet to be sanitized. It disinfects rooms with an incredible speed, finishing the task in about four minutes. An interesting piece of technology called the Helios UV disinfection robot, eliminates the virus strain with limited amounts of UV-C radiation at a time to avoid adverse effects on human health (Sascha, 2020). These technologies will not only quickly reduce the threat of the virus in hospital settings, but will also reduce the chances of patients and staff from acquiring secondary infections or diseases such as skin cancer, associated with prolonged UV radiation exposure (American Cancer Society, 2020).

A part of preventing infection also includes detecting and sanitizing viral hot-spots (see Chapter 7). With the pandemic in full force, different kinds of robotics and AI have been designed and installed. Some of these include using robots which track all COVID-19 outbreaks through wastewater. The robot quickly maps the results which can be used to identify hot-spots that must be dealt with immediately (Chakraborty, 2020). Another includes a human-resembling, or humanoid robot named Pepper, which dispenses disinfectant solution in public settings and provides information to users on how to stay protected from the coronavirus. The wastewater method deals with detecting the virus, which it does with a high degree of accuracy, while Pepper ensures that individuals take the sanitation precautions necessary in large public settings such as hospitals, airports, and shopping centres, which are prone to high cases of infection (Atalayar, 2020).

Nurses have been hard-hit by the pandemic. They have been at the frontline working themselves to the bone, serving patients for an innumerable number of hours. Robotic technologies such as Tommy the Robot Nurse and Cloud Ginger may help lessen the responsibilities that nurses have to fulfill when providing care to coronavirus patients (see Chapter 8). Tommy measures blood pressure and oxygen saturation in patients which is part of the standard nursing protocol for coronavirus patients. Cloud Ginger helps with administrative duties and has also been used to reduce the anxiety levels of patients and staff in hospitals. Both technologies reduce the extensive need for hospital staff and PPE while ensuring that patients are taken care of in an environment that contains both physical and mental aid facilities.

Apart from nurses, the pharmaceutical industry has also been suffering as a result of the pandemic. We are seeing manufacturing delays and drug shortages, both of which are taking a toll on public health and wellbeing (see Chapter 9). To overcome these challenges, India's Cipla manufacturing plant has created a new execution system which automates material flow and associated costs (Banga, 2020). It includes an up-to-date digital record of all products on the assembly line, which ensures that they meet quality standards upon delivery. Likewise, Japan's Musashi AI has hired its first two robotic employees which include a Visual Quality Control Inspector and a Fully Autonomous Forklift Driver. Using AI, they identify any errors in medicine manufacture and ensure that products are stored and delivered appropriately and timely (Fretty, 2020). Incorporating such technologies into production lines can ensure that drugs and healthcare products are released efficiently and effectively.

Although these robotic technologies have been successful in facilitating healthcare procedures, it is important to consider the cost and time efficiency, security, as well as economic and psychological impacts that they could have which may affect our ability to cope with and recover from the pandemic in the long-term.

Without a doubt, AI and robotics have helped medical staff and administration save time and money. For instance, excellent COVID-19 tests can be administered and disinfection can be performed in small amounts of time. This allows for healthcare workers to dedicate more of their time towards duties which may require their skill sets and not just labour. Such technologies also guarantee safety, reduce chances of human error, and risk of infection all of which could have large financial consequences on healthcare businesses. As mentioned in Chapter 10, although some of these technologies have high initial and maintenance costs, the efficiencies they grant us make them worth investing into.

We must consider the entire economic picture however, to determine whether robotic technologies would be truly worth investing in. As mentioned in Chapter 11, healthcare automation would increase productivity and growth since tasks can be done quickly and more efficiently. Machines can take care of duties which require lots of manual labour and ensure that they are conducted in a safe manner. The downside is that some employees, especially those that are lowly-skilled, can be displaced, which increases national unemployment rates. This can ultimately drive us into a deep recession unless new opportunities are created. On the other hand, integrating AI does open the doors for high-skilled workers in the quaternary sector, such as those who work in IT and technology-based fields. Healthcare businesses would have to keep in mind the costs associated with training prior or new workers to use advanced machinery.

Being made redundant, the inability to find a well-paying job, and financial hardship associated with the pandemic will have lasting psychological impacts on our citizens. Chapter 12 investigates how the effect of restricted business operations and lack of social interaction due to quarantine will impact different age groups. The percentage of the total population that has faced job losses for example, has seen a greater number of negative mental health indicators including exhaustion, burnout, and stress (Patel, Devaraj, Hicks, & Wornell, 2018). Prior to automating most healthcare procedures, public mental health must be weighed as a factor.

Lastly, a concern that we all have when using technology -- privacy and security. Healthcare automation requires the collection and storage of patient data which is highly sensitive. There is the possibility that your data could be used for an alternative purpose without consent or medical agencies could be hacked, as discussed in Chapter 13. This isn't meant to discourage countries from implementing machines in healthcare systems, but rather, are risks to be aware of when providing personal information.

COVID-19 is one of the largest pandemics that we have ever come across in history. Modern-day healthcare systems have been taken down and economies have begun to decline due to the spread of the virus. To bring the world back into recovery, automated processes are needed, even with their many drawbacks. We hope that by shedding light on technologies being developed and employed in different healthcare systems, countries will be inspired to implement such solutions and perhaps come up with new ones to address the challenges associated with COVID-19. The conditions that the virus has created are different worldwide, but we can all agree upon one thing: this virus is nothing like what we have seen before. It's much more advanced. And so to overcome an advanced virus such as COVID-19, we have to take a step further from the traditional methods that we are so accustomed to, and advance our healthcare systems with artificial intelligence applications.

COVID 19 and the need for automation

Imagine living a hundred years ago from today. There would be no washing machines, coffee makers, or air conditioners. We encounter so many forms of machinery in our daily lives that we forget even the simplest parts of our everyday tasks are automated. This goes to show how rapidly we have advanced in this field over the years. However, it is possible that this growth rate may speed up even more in response to a pandemic.

COVID-19 can be a turning point, or perhaps the wake-up call, needed for the rapid enlargement of the automation industry. In Canada, we saw a swift lockdown of the entire nation in just days with schools, workplaces, and essential business services all coming to a halt. As a nation, we were simply not prepared for the virus. This was not just the case for Canada, but also many other countries around the world such as China, India, and Spain, which are only some of the countries with a higher number of cases.

Despite lockdown restrictions, essential businesses and services quickly adapted to the new norm. Classrooms were now all online, workplace meetings were held on Zoom, and retailers were open for business through online shopping. Even though these practices may help us feel as though we are reaching a closer-to-normal lifestyle today, it does not go unnoticed that automated services remain perfectly functioning. As a matter of fact, automated services are much safer during these times than some of these online services. This makes sense because robots and other forms of automated machinery cannot get infected with the virus. Ordering online can require human-to-human contact to manufacture, process, and deliver the goods. No matter how many safe practices, such as social distancing and wearing protective equipment, are put in place, there will always be a risk to humans.

Another industry that has emphasized the need for automation due to COVID-19 is healthcare. Simply put, the healthcare industry has undergone major changes in terms of its procedures and services, and rightfully so because non-essential businesses can come to a pause, but healthcare is, and will always be, extremely important. Everyday, there are millions of frontline workers putting their lives at risk to not only help treat those infected with the virus, but also to continue providing healthcare to those in need. Automated systems can help to implement safer practices to both the frontline workers and the patients.

As a result of the pandemic, we are now realizing the severe need for automation in other industries as well, ranging from its enterprise use to the use of robots and artificial intelligence in healthcare. There can always be contingency plans for natural disasters, but there simply cannot be a plan for pandemics without including automation in the picture. This chapter will focus on this need for automated procedures by looking specifically at the use of automation in healthcare and related fields.

A large part of the healthcare sector is the business industry, as a number of health-related services depend on businesses. Therefore, the consequences that we saw in regards to the impact of COVID-19 on essential businesses ultimately had an effect on healthcare services. The main impact of this is that consumers are not receiving the goods and services that they need because businesses are not operating. This may not seem that significant on the surface, however, it is often forgotten how some vital products have been affected due to this.

An example of necessary products comes from the pharmaceutical industry. Even though this industry has always seemed to be very slow in adapting their processes in response to changing times, pharma operations were able to quickly adjust to accommodate for the COVID-19 changes. Nevertheless, this had to be done because patients needed their medical treatments and often, these medicines would travel across borders. Additionally, workplace safety had to be met for both employees and employers, which was another challenge that was faced by this industry. For these reasons, automation has been shown to be important in response to COVID-19.

One major change that many pharmaceutical distributors had to undergo was to figure out a safe, yet effective, way of reaching their customers to deliver their medicines. OmniSYS, an American technology company, has recently begun offering free services to pharmaceuticals in response to the COVID-19 crisis. They help communicate with patients by using interactive voice response (IVR), which is an automated voice messaging system that can convey customized messages to patients (OmniSYS, 2020). From this example, we can see how important automated systems are necessary in this time in order to continue providing essential services.

In addition to these changes to help cope with the impacts of COVID-19, there has been a need for automated procedures to directly help against the fight of the virus as well. Surely, this is where the main benefits of automated systems are seen, as they allow for advances closer to the treatment for the virus without putting more lives at risk. One major issue that is being faced on a global scale is the lack of personal protective equipment (PPE). This is occurring because of increased demand, panic buying, and misuse (WHO, 2020). The insufficiency of this equipment is contributing to the rise in the number of cases, as frontline workers in the medical field have to sometimes reuse their PPE for days, making them more susceptible to the virus.

According to the World Health Organization (WHO), in order to overcome the insufficient PPE, manufacturing must be increased by at least 40 percent (WHO, 2020). As a result, many well known businesses are ramping up their manufacturing abilities using automation. One of the most notable is General Motors (GM), a vehicle manufacturer. They partnered with a United States based automation company to build an assembly line that would automate the process of producing face masks. This line was built in less than a week and has the capacity to produce 50,000 more face masks per day (GM Corporate Newsroom, 2020). These masks are then deployed to areas in critical need to help combat the virus. Once again, this example emphasized the need for automation, as this increase in manufacturing directly helps fight the virus.

In addition to the need for protecting frontline workers from those that are infected with COVID-19, automation may also be beneficial to test individuals to see if they have the virus. In this way, a significant burden would be lifted off of these workers, as they can then help focus on treating those infected and ensuring their recovery, rather than facilitating testing centres. It also can ease tensions about the risk of getting the virus at these centres, which may motivate more people to get tested.

Furthermore, if the process of physically testing people can become automated, then the procedure of processing test results and reporting these can also benefit from automation. This would lead to faster results, which is important because the quicker the virus is diagnosed, the quicker the infected individual can take the appropriate isolation measures. Next, automated processes can be more accurate than humans, as they avoid common human errors such as accidentally recording false information. Additionally, faster processing and greater accuracy will allow professionals to receive a clear picture on the current situation in terms of the number of cases and the frequency of testing. This information will allow a particular province or state to determine if their prevention methods are effective and what the next steps should be in a timely manner.

Finally, another area that has been a particular focus in North America is the long-term and nursing homes crisis. These homes have become a hot-spot for COVID-19 infections for many reasons. One reason is that the patients in long-term care are extremely vulnerable and therefore are susceptible to the virus. To add onto this, they do not have much physical space, which is a result of the differences between the design of these homes in comparison to a hospital. Furthermore, they are generally under-prepared and do not have access to the appropriate PPE and personnel that are usually put in place in terms of a crisis. Unfortunately, the result of this during the pandemic is a high mortality rate in such homes and the lack of family interaction. This situation highlights the need for automation, as workers in these homes are overwhelmed and highly exhausted. Automated robots can help with parts of their jobs, such as communicating with the patients. Another system that has been proposed to help with this issue is the use of Global Smart Home Automation systems. This primarily includes smart home devices which help to control systems such as lighting and security (MarketWatch, 2020). Implementing home automation like this will greatly reduce the workload faced by workers, which is especially relevant to consider during a pandemic.

However, it is important to note that even though COVID-19 has shown the need of automation, it cannot be implemented without facing challenges. To begin, the simple lack of human interaction is one example of this. Many studies have shown the positive impact of human interaction on mental and physical health. For example, having social connections reduces stress, as we have people around us to share our lives with. Additionally, having co-workers to collaborate with in the workforce can generate meaningful relationships and communities, which further helps us to feel socially connected and engaged. This is particularly important to consider in the healthcare industry, as medical services can lose this human touch which might have a negative impact on patient health. Therefore, automation should not completely replace human interaction, but complement it in an appropriate balance.

Another aspect of automation that needs to be considered is the risk of cyber attacks. In the healthcare field, there is a lot of highly sensitive and private data that should be well secured in order to maintain the quality of healthcare and the trust of patients. However, to prevent any such event from happening, it is important to ensure that automated systems are extremely well secured and managed if they were to be implemented in places such as hospitals and clinics.

Furthermore, the next main factor to consider with the need for automation is the degree of competence of automated systems. If automation becomes integrated with direct patient care, these systems must be highly proficient in their roles. For example, if the process of giving patients medicine becomes automated, then the robot performing this task has to be accurate with the dosage and timings. The trust of patients on healthcare systems is dependent on factors like this as well and thus, even one mistake in automated processes can influence the extent to which people consult healthcare services.

To conclude the ideas that were detailed in this chapter, we realized the need for automation due to the implications of COVID-19. This need is emphasized during this pandemic because humans may contract the virus, however, robots and other automated devices cannot. Therefore, in order for essential businesses to continue providing their products and services, automation is key to ensure worker and consumer protection. An industry in which the operations are vital is healthcare. These services are important both to treat those infected with the virus but also to continue providing treatment to regular patients. To ensure that these services are continued during world crises, automation is required in the pharmaceutical industry to secure the supply of important medications while implementing safety protocols. Automation has also shown to be beneficial to combat COVID-19 directly. This includes the manufacturing PPE at a higher scale to overcome the global shortage, ways to conduct and process COVID-19 tests, and help with the long-term and nursing homes crisis by easing the stress on frontline workers. Even though the automation of these processes have their respective limitations, it is important to realize that these are just a few of the examples in which automation can help society function during a pandemic.

Robotics in the diagnosis of COVID 19

Introduction

Early into the world's battle against COVID-19, South Korea clearly established themselves to be at the forefront of controlling the pandemic. Although South Korea was one of the hardest hitting countries by the end of February, with more patients than China at the time, they quickly established themselves as one of the most successful countries at handling the outbreak (COVID Tracking Project, 2020). In comparison to other countries, South Korea was able to "flatten the curve" more quickly than the majority of countries - an impressive feat in itself, considering their large population. During an interview in March 2020, Tina Park from the Canadian Centre for the Responsibility to Protect stated that South Korea's success in handling COVID-19 came from the strong collaboration and transparency between the South Korean government and their labs, giving people quicker access to testing, leading to earlier diagnosis (CBC News, 2020). For instance, according to the Atlantic's Covid Tracking Project, from the end of January to the end of April of this year, South Korea tested almost more than four times the number of people than the U.S. did, with South Korea testing 348,395 cases and the U.S. testing a mere 94,514 people (COVID Tracking Project, 2020). Thus, when looking at a country's future ability to succeed against subsequent waves of COVID-19, it is important to take a deeper look into the technologies countries around the world have been using to test and diagnose patients faster with COVID-19 and their associated symptoms. This factor in particular may have some impact on future outbreaks, as countries such as South Korea have proven that rapid nationwide testing is the best approach at handling large outbreaks.

Safety for Frontline Workers

Many countries have realized the huge opportunity automated technology such as robotics can bring towards the health industry. In particular, most researchers find the technologies to be most beneficial during outbreaks such as COVID19 because they offer a method to keep frontline workers safe that other methods would not provide for them. In Denmark, Flow Robotics and scientists have created a pipette robot that is able to test over 1,000 COVID-19 test kits a day, providing quick feedback for the clinicians at both German and Danish hospitals. Using technology, they were able to create this quick pipeline of twelve robots to provide testing and diagnosis for patients in just three weeks (Flow Robotics, 2020). Furthermore, this pipeline is not only more efficient with technology, but also more effective because these robotics minimize the pipetting errors that may be caused when scientists are handling such viruses. They provide protection for many of these scientists, as they do not have to directly come into contact with the virus, decreasing their chances significantly of contracting the virus.

Other researchers have been involved in creating virtual clinics through the technologies of telemedicine consultations, telecommunications, and artificial intelligence (Bahl et al., 2020). These virtual clinics provide consultations for the patients' benefits, while also maintaining the health of physicians. Without telecommunications, this would be difficult to accomplish. The consultations would have required meetings to be in person, increasing the physical crowding of patients in hospitals and clinics and the risk of both patients and doctors of contracting the virus. Telecommunications have also been beneficial for manufacturing industries who have their interestings in biomedical and pharmaceutical companies, as they have increased transparency between both parties and have led to quicker, more efficient, and more safe production lines. With wireless connectivity and sensors, many healthcare production lines that create testing kits for diagnosis have been able to be done remotely through such technologies, keeping many of the workers in those industries safe (Bahl et al., 2020). Additionally, research in both artificial intelligence (AI) and Internet of Things (IoT) fields have provided great benefit, as they allow for real-time processing of information.

This allows physicians to update their patients more quickly and more effectively concerning diagnoses. Technology has become crucial in hospitals' efforts in providing treatment for individuals who live in remote areas. They are able to provide not only remote consultations and treatments, but also diagnostic testing, such as remotely measuring the temperature of a patient (Carriere et al., 2020). Other robots that have been known to do diagnostic testing in research facilities are Med Tech Boston's RP-VITA and the Rush University Medical Center.

During the pandemic, the demand for front-line workers have increased dramatically, with many hospitals and clinics not having enough workers due to the large volume of patients. However, countries such as Belgium have created "receptionist robots" to help guide patients, sending them see appropriate physicians based on the symptoms they describe to be presenting (Chang et al., 2020). The robot has seen great success in Belgium, and it is extremely popular amongst children, as they seem to be a helpful and friendly option for children, especially for challenging and daunting times such as during the pandemic.

The benefits of providing safety for frontline workers is a two-way street. It not only protects the frontline workers, who have worked extended number of hours caring for patients and risking their lives to do so, but it also protects everyone else. Robotics have provided ways to reduce the burden for frontline workers and have given patients more effective and efficient treatment and consultations that would have otherwise not been possible with technology. By far this benefit was the most significant weighting that most healthcare companies had, when considering the use of technology. These technologies have also reduced the volume of patients at emergency room hospitals, reducing the burden put on these systems that may not be able to always handle so many people at once.

Diagnosis Technology

Robotics and technology have made it possible to complete medical tests more efficiently, decreasing wait times for patients. AI has been used to develop online medical examinations for patients, producing CT scans which are required for detecting pneumonia, a major symptom that arises due to the contraction of the virus (Bahl et al., 2020). Researchers from Renmin Hospital of Wuhan University, have been able to create deep learning models that predict the likelihood of an individual contracting COVID-19 with 95% accuracy from CT scans of pneumonia. Taken from a selection of 45,000 CT scans, with only 51 pictures of those who have also contracted COVID-19, this deep learning model has provided significant insights on the possibilities technology will have in the future in diagnosing diseases with more efficiency and effectiveness (Khari, 2020). Similarly, technologies such as 3D scanning, virtual reality, motion capture, and robotic mapping can help with thoracic chest canning, another useful tool for detecting symptoms related to COVID-19 (Bahl et al., 2020). A similar piece of technology are radiologist robots such as Multitom Rax, the twin robotic X-ray from Siemens Healthineers that provides scans for physicians more quickly than previous pieces of technology (Chang et al., 2020).

A technology that has recently been deeply researched by scientists are glucose monitors, a biosensor used in clinical analysis. Although still under development, this wireless biosensor patch is a real-tracking single-use piece of technology that monitors all types of symptoms that could suggest a patient contracting COVID-19, such as temperature, electrocardiogram traces (ECG), and respiration rate (Bahl et al., 2020). Overall, new pieces of technologies have allowed physicians to be more efficient than ever before, ensuring reduced waiting times for patients. There is a big potential in this industry to grow further, and many more technologies will develop to not only make testing and diagnostics more efficient, but also more effective.

Predicting Diagnoses

One of the biggest benefits to using technology to diagnose patients is not about just the patients who come in and voluntarily come to get themselves tested. Technology has allowed us to do this by tracking devices and using computer vision to find people who have been recently in contact with those who have been infected. This decreases the number of diagnoses and treatments and has been a major advantage at reducing the number of cases that are experienced. Furthermore, it allows patients to be held to some accountability to ensure that they are self-quarantining and not affecting other people (Bahl et al., 2020).

Faster Testing

One of the biggest manners that South Korea was able to control the outbreak so quickly was due to their mandatory testing policies, where they directed everyone to be tested. However, not all countries have fallen suit, and many of them are lacking the resources and the personnel to set up these testing centers as quickly as Korea did. There are collaborations that are aiming to make testing and testing kits more available for everyone. For example, a popup laboratory at University of California Berkeley's Innovative Genomics Institute (IGI) has created robotic pipelines to speed up the number of tests they perform, specifically for those individuals that live near the Bay Area. Partnering with the UC Berkeley Tang Center and medical centers around the Eastern Bay area, in a matter of weeks, they were able to set up a process that enabled from 1,000 to 3,000 tests a day (Sanders, 2020). Normally, these procedures can take from many months to even a few years. Robotics have helped make the Polymerase Chain Reaction (PCR) phase of testing samples more efficient, a process which is essential to amplifying the RNA available for testing. The results have come back extremely quickly as well. Results that would normally have taken from five to seven day to receive the results now only take 12 to 24 hours for the patients to know their results. This shows how crucial robotics technology is becoming, as it has enabled quicker pipelines in the healthcare industry.

Additionally, companies have been looking for ways to speed up the production of testing kits. Two Maine manufacturing companies have increased their production levels to produce over one million testing kits, which is currently greatly needed in the United States to combat COVID-19 (Lanco Integrated, 2020). These testing kits are in high demand. For instance, a Germany company requested for testing kits to be made in eight weeks, showing the high demand and need for testing kits. Without these testing kits, effective diagnostic procedures cannot occur because they would not have the resources to do so. Thus, it is vital for companies and researchers to continue making the production of testing kits as their priority.

Conclusion

Clearly, many technologies go into diagnosing patients, from providing a remote means of consultation to protect both the physicians and the patients to creating scanning software that is more effective and efficient than CT scans. It has also become apparent how crucial technology has become in our efforts against COVID-19, as it reduces our burdens and allows us to focus on goals and tasks that are more important. Robots have been used to automate procedures at hospitals, so that doctors have more time to attend to a greater volume of patients. This has also shown that despite the breakthroughs in technology, it is not enough. Although we have made these testing and diagnostics pipelines more quickly than usual, many countries are still lacking the testing kits and the resources to quickly diagnose people, and this is why the number of cases in most countries continue to increase every day, rather than to flatten the curve. It is important to recognize that using robotics to diagnose patients goes beyond the diagnosis itself, but involves all the steps that lead to the diagnosis, such as manufacturing of test kits or data collection. Most crucially, it is important to look beyond technologies of one specific field in the world's battle against COVID-19. Having a bigger picture of all the technologies and how they work together will give us a picture of the profound impact technology has had, even if it is not enough just yet.

Intelligent thermometry systems

The scene of a child waking up sick and having a parent check their temperature with a thermometer has become cliche in the film industry. Oftentimes reality is similar and when expressing discomfort or the possibility of sickness to friends or family, a common response is to check if the forehead is unusually warm. Taking someone's body temperature is a good indication of telling whether or not they have a fever. Usually caused by an infection, fevers cause a rise in body temperature, because the body is trying to fight off that infection by creating an unsuitable (warm) environment. The average body temperature of an adult is approximately 37°C (98.6°F), but often varies 0.5°-1°C (1°-2°F). However, body temperature is usually lower in the morning, due to being inactive while sleeping, and increases during the day, reaching a high during the late afternoon or evening. When someone has a fever, their temperature is around 38°C (100.4°F) or above, but if it reaches 38.8°C (102°F), one should seek a healthcare professional (Mayo Clinic Staff, 2019).

Measuring body temperature has become increasingly important during the past few months because of the COVID-19 pandemic. One of the most common symptoms of the novel coronavirus is having a fever and because of this many have tried to purchase thermometers to self-diagnose and determine if they potentially have the coronavirus (CDC, 2020). However, the fears associated with staying indoors and quarantining have caused many across the country to stock up on important resources such as toilet paper, hand sanitizer, and thermometers. Additionally, high-traffic areas like hospitals and airports have started to increasingly use infrared body temperature measuring devices to screen people coming in and out of buildings for abnormally high temperatures, as a potential indicator to having the virus. This has brought up many questions regarding the functionality and effectiveness of such intelligent thermometry systems as well as any recent advances in thermometry that may help to combat the pandemic.

Beginning with the basics, thermometry is a thermal analysis technique that usually measures temperature together with time. One of the oldest and simplest forms of analysis, thermometry has been used to plot heating and cooling curves for certain substances and reactions to indicate processes like changes in phase (solid, liquid, gas, plasma). In the scientific perspective and a laboratory setting, the most common temperature measuring tool is the mercury-in-glass thermometer. It's a glass tube marked with the standard temperature scale and filled with liquid mercury. As the temperature rises or falls, the mercury expands or contracts accordingly. The rate of expansion or contraction is calibrated on the glass scale so that an accurate reading can be given. The tiny bulb at the bottom and the micro-fine size of the tube aids the mercury in reaching the temperature of what it's measuring very quickly. Given the poisonous nature of mercury, it should absolutely not be used for measuring a person's body temperature or be put in someone's mouth! Nowadays, mercury-in-glass thermometers have been replaced with digital thermometers for use in science class or actual laboratories (Swenson & Quinn, 2004). Another popular method of measuring temperature that has a lot of industrial applications is infrared thermometry. Although not a viable method for measuring the interior temperature of an object, infrared thermometry is perfect for measuring the surface temperature of an object. It follows a simple principle: everything that has mass emits some amount of energy in the form of heat. By using the difference between the infrared rays coming off of the object and those from the surrounding environment, the surface temperature of an object can be determined. An infrared thermometer focuses light coming from an object in the form of infrared rays and channels that light through a detector, called a thermophile. Within the thermophile, the infrared radiation is turned into heat which is then turned into electricity which is finally measured to give a temperature reading. A great benefit to using an infrared thermometry device is being able to measure temperature from a distance when unable to insert a probe. The thermometry systems mentioned above have many smaller, distinct categories that are applicable to a hospital setting (Transcat, 2018).

In the world of health-care, many different types of thermometers and thermometry devices may be required depending on the patient of interest. The most common are digital thermometers. They're fast, the most accurate, and take reading from under the tongue, the rectum, or under the armpit. With regards to those specific areas, the armpit is least accurate while rectal is best for infants from three months to three years old. A digital thermometer contains a small computing mechanism and a resistor that act as electronic heat sensors. When there is a change in temperature, the sensor measures a change in resistance, and the computer converts the change in resistance to a change in temperature. It should be noted that users should wait fifteen minutes after eating or drinking to take an oral temperature, as not doing so could provide inaccurate readings (Mayo Clinic Staff, 2018).

Another common one used in walk-in-clinics is the electronic ear thermometer (tympanic thermometer). By using infrared technology, these thermometers can measure the temperature inside the ear canal, but readings may be inaccurate depending on how much wax is present in the ear. As opposed to a digital thermometer, an electronic ear thermometer may be easier to use on children and babies, as it is difficult to get them to sit still. In terms of a specific age range, they are good to use for infants older than six months, older children, and adults. Going back to feeling the forehead for abnormal warmth, forehead thermometers, like electronic ear thermometers, use infrared technology to measure surface temperature. For appropriate use, they are placed on the temporal artery and are less accurate than digital thermometers. If there are no forehead thermometers present, a plastic strip thermometer is a likely alternative. Although it doesn't give an exact temperature reading, placing the strip on the forehead can help to detect if the individual has a fever. It should be noted that the plastic strip thermometer only acts as indication that something is wrong and doesn't give any definitive answers. A pacifier thermometer is great for using on babies that are older than three months, but requires them to sit still for a few minutes. The baby sucks on the pacifier until the temperature is recorded, but this reading may be inaccurate if the child cannot sit still for a short period of time (Mayo Clinic Staff, 2018).

A final note should be taken in regards to cleaning and maintenance. The tips of digital thermometers (and most other ones) should always be cleaned with soap, water, or alcohol before and after use. The cleaned area should also be rinsed with lukewarm water, and afterwards, the device should be stored in a dry area that's not exposed to drastic changes in temperature. If there are multiple thermometers, they should be arranged in a way that one does not confuse the rectal thermometer for the oral one or vice versa (Mayo Clinic Staff, 2018).

With the continuous breakthroughs in technology that occur every year, many are currently trying to implement these ideas into bettering healthcare and diagnosis. A notable example, although not completely related to the COVID-19 pandemic, is coupling artificial intelligence and infrared thermography. A study published in February of 2020, tried to find an objective, fast, and cost-effective method for automatically identifying different stages of cellulite using infrared imaging. Cellulite is a common physiological condition experienced by 85-98% of post-pubertal females, where the skin around the hip, thighs, and buttocks area adopts a puckered and dimpled appearance. The study used artificial intelligence to develop a custom image preprocessing algorithm to automatically detect cellulite regions. 212 female volunteers between the ages of 19 and 22 participated in the study, and by using an artificial neural network, all stages of cellulite could be discerned with an average accuracy higher than 80%, with primary stages having an average accuracy higher than 90%. The study concluded that implementing a computer-aided automatic cellulite severity identification device that used infrared imaging was feasible and reliable for diagnosis. It's important to pay attention to studies like these, as they may have the potential to be retrofitted or modified to solve other problems like those associated with the COVID-19 pandemic (Bauer et al., 2020).

As cases started to rise and the world was brought to its knees by the COVID-19 pandemic, a lot of concerns about thermometers and other intelligent thermometry systems were brought to light. By looking at both news articles related to thermometers from March to April of 2020 as well as documents mentioning thermometry devices from national organizations, one can analyze the similarities and differences between the two in order to figure out if there were any societal gaps in knowledge or confusion during the time.

On February 16, 2020, Forbes published an article questioning the functionality of thermometer guns in regards to diagnosis of COVID-19 related symptoms. It was mentioned that there was an increasing use of infrared body temperature measuring devices at high traffic areas like airports and health care facilities. The article argued that body temperature is not necessarily a good indication of whether or not one has been affected by the coronavirus. Specifically, although a fever occurs because of the body's immune response to foreign invaders (virus), many people may be taking medications, which could lower the increased body temperature to normal. Similarly, a person's immune system may have not realized that there are foreign invaders, meaning the person could be infected, but their body temperature remains normal. All these instances could result in thermal screening devices not recognizing those who are actually infected and falsely recognizing those who may not be (Lee, 2020).

During the next month, on March 31, 2020, the CBC (Canadian Broadcasting Corporation) published an article about frustrating thermometer shortages in the city of London, Ontario. Many pharmacists noted that the outbreak had created such a high demand that there was no supply of thermometers left for at least a few months. Even the more expensive (~$50) digital thermometers were very low in stock and quickly running out. Health professionals and pharmacists voiced their concerns about potentially sick and symptomatic individuals going from one store to the next, in pursuit of a thermometer. They advised people to stay inside and isolate themselves as much as possible, and even more so if they were showing symptoms, as cases were increasing at a very alarming rate (Lupton, 2020).

Lastly, on March 30, 2020, The New York Times published an article about fever and its association with the coronavirus. It brought up the problem that although the CDC recommends potentially exposed individuals to take their temperature twice daily, most young millennials don't own any thermometers. Additionally, normal temperature numbers vary from person to person and depend on factors like height, weight, gender, weather, etc. Since most people are cooler in the morning, the second temperature reading should be taken sometime in the evening. Dr. Brad Uren, assistant professor and emergency doctor and the University of Michigan, says that it's not necessary for people to purchase expensive, high-end thermometer devices and that a cheap one will suffice (Luthra, 2020).

With regards to exciting advancements, on April 2, 2020, the CNBC (Consumer News Business Channel) reported a company that created a smart thermometer that could aid in detecting COVID-19 hotspots potentially faster than the CDC (Center for Disease Control and Prevention). The San Francisco based health and technology company, Kinsa created a internet-connected smart thermometer that pairs with a mobile app. They are a little bit pricey, with the standard model being around $40 and $70 for the ear model. CEO and founder of Kinsa, Inder Singh stated that there are more than 2 million users across the United States and about 150 thousand temperature readings per day. Users upload temperatures and symptoms to Kinsa's database via bluetooth and it aggregates all data nationwide, providing an abundance of data for the company to then analyze. The database contains previously collected data from a few years back that can be used to determine what a normal cold and flu season looks like. Kinsa uses a compartmental model that tracks the proportion of the population that is susceptible and infected over time. Then it looks at the rate at which individuals recover from the infected state and the rate at which susceptible individuals transition to becoming infected. The compartmental model is extremely versatile, initially being used in the 19th century to map chemical reactions, but now utilized for epidemiological modelling with modern computing. The model is not too complex either, and can be simplified down to about 50 lines of code. Utilizing the model, Kinsa can predict the next flu outbreak approximately 20 weeks in advance, which allows for the health industry to better allocate its resources to the areas that need them the most at the right times. It should be made clear that when a spike is seen in the system, it does not definitively mean that COVID-19 is present in that area, but that testing kits should definitely be sent there. In comparison, the CDC relies on people to visit healthcare facilities, which then have to make diagnoses and then report those diagnoses. This creates a lag time that Kinsa doesn't have, as it's tracking everything in real time (Indelicato, 2020).

Now, the information given in these articles can be compared with the information found in two thermometry related documents from national organizations in Canada and the U.S. On May 6, 2020, the CADTH (Canadian Agency for Drugs and Technologies in Health) released an evidence assessment on infrared temperature devices for COVID-19 screening during the outbreaks. The research questions included: what is the evidence on the effectiveness and accuracy of non-contact thermal temperature/imaging screening systems for screening populations, what are guidelines on the use of non-contact thermal temperature/imaging screening systems, and what are the risks and safety guidance compared to conventional approaches to temperature screening. The objective of the assessment was to review literature on the accuracy of infrared temperature screening devices for identifying visitors or staff that entered healthcare buildings who may have a potentially infectious illness. It was concluded that there was insufficient evidence to support that non-contact infrared temperature screening methods were effective for detecting infected individuals. Thus, non-contact infrared temperature screening devices may not be necessary for screening potentially infected staff or visitors who enter certain facilities. The specific limitations include generally low or inconsistent sensitivity for the devices examined. Some factors that affect the performance of these devices include but are not limited to: environmental temperature, use of medications that suppress a fever or elevate body temperature, and physical activity. There were also clinical characteristics regarding the coronavirus itself that may affect the accuracy of temperature screening. These include: the variation in the periods of incubation and time-to-onset of clinical symptoms and the proportion of infected individuals who have no fever, mild symptoms, or no symptoms at all who won't be detected through screening (CADTH, 2020).

Moving down south, on June 19, 2020, the FDA (U.S. Food and Drug Association) released information regarding non-contact temperature assessment devices in relation to COVID-19. It states that even when using the devices correctly, temperature assessment may have limited impact on reducing the spread of COVID-19 cases. Some studies suggest that temperature measurements by themselves may miss more than 50% of infected people. Much like the news articles and CADTH assessment, the FDA mentions the possibility of contagious individuals that show no symptoms or elevated body temperature. It is also made clear that an elevated body temperature doesn't definitively indicate that someone is infected with COVID-19, as more evaluation and diagnostic testing is necessary. Non-contact temperature assessment devices do hold some benefits which include: quick temperature readings, a large number of people being evaluated individually at different entrances, and minimal cleaning. However, there still exist limitations and are very similar to the ones listed in the

CADTH assessment. There is also information given in regards to manufacturer labelling and instructions for use. It states that the specific instructions and additional information provided with each thermometry device should be read to increase accuracy and that temperature measurement may be affected by clothing and the training of the person handling the device. The FDA has also issued guidelines/enforcement policies to increase and expand the availability of thermal imaging systems and clinical electronic thermometers (Center for Devices and Radiological Health, 2020). In summary, taking body temperature is an important indication of human health and should always be done when feeling ill. Thermometry is an effective analysis technique used in the healthcare industry, but because of the COVID-19 pandemic, the supply of thermometry devices has been put under constant strain. As restrictions ease up, those who feel that they aren't showing any symptoms can go out and purchase thermometer devices if necessary, but should follow proper safety precautions (wearing mask and gloves). There have been many advancements in thermometry over the years, some with artificial intelligence that may potentially be applicable to COVID-19 and other diseases. When comparing a small group of news articles and documents from national organizations across Canada and the U.S., much of the information presented in the two mediums were quite similar in regards to thermometry devices, suggesting small knowledge gaps in society.

Robotics in Sample Collection

The COVID-19 pandemic is arguably more lethal than the seasonal flu. It is one that has challenged and exhausted our healthcare systems. It has spread from country to country even in the smallest of towns, making it difficult to keep track of. The number of cases that we see reported on the news on a daily basis are not an accurate representation of the numbers seen in reality. This is mainly due to our inability to administer enough tests, which are often conducted by healthcare professionals. As a result, there has been an increased demand for new technologies and robotics to ease sample collection and testing procedures in order to correctly diagnose and report coronavirus patient numbers.

As of 2020, two types of tests have been approved to diagnose COVID-19. These include viral and antibody tests, with the viral test being the most commonly administered. In this test, the healthcare worker collects fluid from a nasal or throat swab or can collect saliva samples for analysis onsite or in an outside lab (Mayo Clinic, 2020). While performing these tests, the administrator is required to take additional precautions to minimize the risk of becoming infected due to possible exposure to respiratory droplets. The new antibody or serology test, is used to determine if one has had an infection in the past (BC Centre for Disease Control, 2020). It requires the healthcare worker to take a blood sample from patients and analyze it for COVID-19 antibodies. Similar to the viral test, coming in close contact with potential COVID-19 patients could put healthcare workers at risk of contracting the virus. It is important that we reduce possible transmission especially to nurses, physicians, and lab workers as they have the potential to spread the disease in a much higher capacity.

To overcome the challenges involved with COVID-19 sample

collection, multiple technologies and innovations have been developed and are currently underway. To conduct viral tests, a throat swabbing robot has been developed by a company called Lifeline Robotics in Denmark. After a patient has scanned their identification card, the robot picks up a clean swab using two universal robot UR3 cobot arms and a custom 3D-printed end effector (Crowe, 2020). The robot prototype then uses the cotton swab-resembling tip to take a sample from the inside of a patient's mouth which takes roughly 20-25 seconds. The robot uses its advanced vision system to locate points to swab inside the patient's throat. Once the swab is complete, the robot will place the sample inside of a jar and screw on the lid to prevent the risk of spreading any respiratory droplets. The jar is then safely sent to a lab for analysis. Such a machine has the ability to revolutionize sample collection. It can scale and provide high-quality test results, thereby reducing the burden on our healthcare systems. In fact, this invention has already been put through clinical trials. The first to be tested was Professor Thiusius Rajeeth Savarimuthu of SDU Robotics who claimed that he "was surprised at how softly the robot managed to land the swab at the spot in the throat where it was supposed to hit, so it was a huge success" (Dalgaard, 2020). This device has been put in a few healthcare facilities and will also be placed in border control offices and airports. Scientists are looking to add nose swabbing capabilities in the system to conduct a full nasopharyngeal test, which will predict coronavirus results with a higher degree of accuracy.

In addition, a robotic liquid-handling system is currently being developed within the city of Hamilton. These "robot colleagues" will help increase coronavirus testing. A team of molecular biologists at the research institute of St. Joseph's Hospital aim to create a technology which confirms test results in just two and a half hours. The team has already designed a highly sensitive test which detects a part of the virus which is less likely to mutate. This ensures that test results are consistent, especially if the virus begins to change its genetic code over a period of time. This test is also very effective at distinguishing between the novel coronavirus (SARS-CoV-2) and other coronaviruses.

Since SARS-CoV-2 can be successfully isolated and detected using this test, the goal now is to increase the volume and capacity of testing. The robotics system prototype that has been created involves a two-stage process in which COVID-19 samples are placed inside of a biosafety cabinet. A part of the apparatus is inserted into each of the test tubes and used to remove an inactive version of the sample for lab technicians to use. Once the samples are inactivated and safe to use, another instrument is used to pick up those samples and process them. The whole apparatus is able to collect and process about 96 samples in 90 minutes (Mitchell, 2020). Dr. David Bulir from the Research Institute of St. Joe's Hamilton believes that "there's so many samples that people can't keep up with it, and so it's a way to address the problem, to be able to do it efficiently and safely." As of July 2020, 700 coronavirus samples are collected and tested per 24 hours in the city of Hamilton. With consistent updates, the hope is to jump from collecting and testing 700 samples per 24 hours to about 1,500 within the same timeframe. Researchers predict that by optimizing the system, the number of tests conducted within the city could rise to 3,000 per day. By creating a compound device which handles and processes samples, turnaround times can be reduced, allowing for coronavirus test results to be delivered at a much faster rate.

To similarly address challenges such as shortage of tests and long wait times for results, Bright Machines has developed a COVID-19 test machine in collaboration with Ichilov Hospital Laboratory in Israel. This automated system consists of a robotic cell, dual-conveyor, robotic arm, and a vision system. The robot is programmed to open different-sized test tubes to draw patient samples and transfer them to control tubes for lab technicians to analyze. This process reduces any redundant and risky manual labour, decreasing the risk of contracting the virus. The vision system installed in the device is used to verify performance unlike in traditional testing, where humans are bound to make mistakes which can lead to further detriment (Rayome, 2020). The automated system from Bright Machines also allows for sample collection and testing to occur 24/7, which has increased the volume of tests completed per day. This machine has not only increased the number and quality of coronavirus samples and tests completed per day, but it has also reduced the need for human involvement which keeps lab workers from contracting the virus.

All of the above robotic inventions have been used to perform viral tests. A medical device established early this year performs blood draws using an automated process, to determine whether a patient has COVID-19 antibodies. The system combines an image-guided venipuncture robot which handles samples and a centrifuge-based blood analyzer, used to make quantitative hematology assessments. Venipuncture, which requires a healthcare worker to insert a needle into a vein to obtain a blood sample or to perform IV therapy, is a universal method of diagnostic blood sampling and testing (Science Daily, 2020). After conducting this procedure, samples are transported to a laboratory for analysis, with results returned to medical staff in a few weeks time. This can be an extensive process since samples are regularly handled and processed by many intermediaries before results finally reach the patient. In the context of COVID-19, this can sound quite worrisome, since the aim is to increase the amount of tests conducted per day to track the spread of the virus.

In addition, there is a high probability of delay in results due to difficulties performing blood draws. These delays are quite common in vulnerable members of the population such as children, elderly, chronically-ill and obese individuals. In fact, clinicians fail to perform proper venipuncture in 27% of patients without visible veins, 40% of patients without palpable veins and 60% of emaciated patients (Science Daily, 2020). The inability to access veins can increase procedure times by up to an hour, raising staff and equipment costs. The delays in processing can also reduce specimen quality and produce erroneous test results, increasing turnaround times. Repeated puncturing of the veins can increase the likelihood of phlebitis, thrombosis, and infections and may even require healthcare workers to target large veins or arteries in the body at a much greater cost and risk (Balter et al., 2018).

The aim is to develop a technology which reduces turnaround times and increases the success of venous access. The venipuncture robot developed has an ultrasound system that allows for it to successfully detect veins. Its centrifuge system allows for rapid analysis. The components of the device ensure that the venipuncture procedure is conducted according to health and safety guidelines and that the samples are analyzed as soon as possible to eliminate the possible threat of specimen degradation. By using such a device, turnaround times are reduced and hospital work-flow is eased, allowing practitioners to devote more time to treating coronavirus patients instead of testing them. According to Josh Leipheimer, a student at the School of Engineering at Rutgers University in New Brunswick, "a device such as this could help clinicians get blood samples quickly, safely, and reliably, preventing unnecessary complications and pain in patients from multiple needle insertion attempts" (Science Daily, 2020). With the combination of robotic and microfluidic systems, the technology combines the accuracy of traditional laboratory testing with the speed and convenience of point of care testing. Researchers are currently working on ways to refine the device to improve success rates in patients with difficult veins to access. With just enough success, this device could revolutionize the way that we conduct other procedures such as IV catheterization, central venous access, dialysis, and placing arterial lines.

We are currently in an era of innovation. Multiple technologies have been invented to reduce the risk that healthcare workers may be predisposed to by administering tests on potential COVID-19 patients. Some of these technologies include a throat swabbing robot which reduces the need for healthcare workers to physically test and handle samples, making them less prone to infection. Another includes a robotics system which extracts an inactive form of the sample and hands it over to lab technicians for testing which can also reduce physical contact with the virus. The robot in Israel opens and draws samples from patient test tubes for testing purposes as well. Lastly, the venipuncture robot draws blood samples and analyzes them right away, reducing the need for clinician-patient interactions which could put the clinician at risk as well as result in complications to the patient due to inaccurate needle insertion. With inventions like these and many others on the way, we can reduce the risk of transmission and successfully eradicate the threat of COVID-19 in our countries.

Robotic Technologies for Infection Prevention

Introduction

When fighting infectious diseases, one of the biggest concerns is keeping frontline workers safe, as they are most vulnerable when treating infected patients. One of the most crucial methods to protect them and decrease the likelihood of infection is the thorough sanitization of all surfaces, medical equipment, and rooms. Even before COVID-19, robotic technology was used to automate sanitization, a form of infection prevention, but the need for such robotics have increased since the start of the pandemic due to the increase in the number of patients. Finding effective robotic technology for sanitizing is crucial not only for the safety of physicians and other front line workers, but may also benefit patients, as this allows doctors more time to care for patients. The robotics that exists for infection prevention mainly focuses on ultraviolet (UV) light and has already been tested and used in many countries in their efforts against COVID-19. With this field of automated healthcare recently booming, it is vital to analyze the issues that surround selecting an appropriate UV-C light technology, as well as understanding its benefits and detriments.

UV-C Light Technology

Almost all robotic technology uses UV-C light technology to kill bacteria and other microorganisms. Of all three of the UV light categories, including UV-A and UV-B, UV-C has the shortest wavelength and is the only category of UV light that has the ability to alter these microorganism's DNA, making it impossible for these microorganisms to function or reproduce, effectively killing them (Phone Soap, 2020). Although most UV-C light is filtered by our earth's atmosphere, scientists have managed to create this wavelength of UV light synthetically, so that they can be used more effectively in the sanitization industry (Phone Soap, 2020). In recent years, many robotic technologies have used UV-C light to develop technologies that target specific problems regarding the sanitization process.

The HELIOS UV disinfection robot limits the UV-C radiation by using two modes, combining its UV-C light disinfection with air circulation (Sascha, 2020). Using the first mode, air will flow into the UV sterilization channels where internally, the UV light will kill the microorganisms found in the air, limiting the UV light emitted from the robot. However, it also does have a second mode that behaves like most sanitization UV robots, emitting condensed UV light to kill the bacteria, as the emitted UV light decomposes their DNA structures.

On the other hand, the UVD robot uses mapping technology through its built-in lidar sensors to minimize the UV light emitted. Using the lidar sensors, through scanning the environment, the robot is able to map the places in a room that needs to be disinfected, identifying which places need the UV light to be emitted to destroy bacteria and which places can be left alone. Despite limiting selecting where to disinfect, the robot is still quite effective, killing approximately 99.9% of bacteria found in the room and only taking between ten to fifteen minutes to disinfect one room (Sascha, 2020).

Another robot, THOR UVC, uses similar mapping technologies, but aims to tackle efficiency. Using its mapping technology, the THOR UVC robot determines the optimal location in a hospital room to be initially placed, which would enable it to deliver the fastest time to disinfect through continuous UV-C energy being emitted. In contrast to the UVD robot, this robot takes four minutes to disinfect a room on average (Sascha, 2020).

Memphis-based TRU-D SmartUVC has been another robot that has been frequently used in hospitals, due to its ability to take into consideration a myriad of factors including, pathogen size, shape, the number of doors, and blinds. Through knowing these variables, TRU-D SmartUVC is able to calculate the level of UV-C lighting that needs to be emitted to effectively for each room. Additionally, the robot includes real-time reporting and analytics that enable workers to monitor the sanitization process, its efficiency, and effectiveness (Sascha, 2020).

Although most of these robotic technologies have been used to clean hospital rooms, the San Antonio based Xenex Lightstrike robot has been specifically designed to target hospital equipment, as the smaller crevices in hospital equipment and tools make it difficult for the other technology to disinfect efficiently. The Texas Biomedical Research Institute used this robot to sanitize their hospital equipment. The robot had the same effectiveness as other robots, killing 99.99% of the microorganisms found, while reporting a higher efficiency given its smaller surface area, cleaning the equipment in less than two minutes (Sascha, 2020).

All these robotic technologies help minimize the risk of COVID-19's exposure to frontline workers and each responds to a specific challenge of using robotics for sanitization and infection prevention. Coronaviruses have been known to stay on inanimate surfaces, such as metal, glass, or plastic, for as many as two weeks, making it extremely imperative that proper sanitation measures are taken to prevent further infection (Collins et al., 2020). These technologies will be vital to implement, as they have shown to be more efficient and effective compared to traditional sanitization methods, and will prevent many unnecessary infections to our frontline workers.

The extent these technologies reduce risk

Although these technologies clearly more efficiently sanitize hospital rooms than traditional procedures, it is still important to recognize the extent that these robotic technologies are effective. Many studies have been written to describe these sanitization technologies' effectiveness, with University of California San Diego's computer science and engineering professor stating that "for disease prevention, robot-controlled noncontact ultraviolet surface disinfection has already been used because COVID-19 spreads not only from person to person via close contact respiratory droplet transfer but also via contaminated surfaces" (University of California San Diego, 2020). It is unquestionable that these technologies should be reverted to, rather than looking at manual disinfection, as these robotic technologies decrease exposure risk to cleaning personnel and could ultimately be more cost-effective and faster.

However, despite this effectiveness, it should be understood that it only mitigates risk and does not remove it. Personnel still have to clean the robots and procedures to use these robots still have to be implemented. Improper cleaning of these robots could affect the ability to sanitize a room in the future and may lead other personnel who clean the robots to be vulnerable to these infectious diseases. Human error in using these technologies where a room is not cleaned properly will also lead to these similar consequences. Although without a doubt these technologies reduce the risk of physicians being infected, there are still factors and measures that need to be taken into consideration before implementation in order for these automated technologies to be as effective as possible.

Potential possibilities for where these technologies will go next

Despite the UV-C light technology being very promising for scientists, it is also important to see what other directions this technology could be taken in to develop more effective technology or more innovative methods that targets a different aspect about sanitization. For instance, in the future, opportunities lie in developing intelligent navigation systems within the robots to detect high risk, high touch areas and determine the optimal prevention measure that needs to be taken for each area (Khan et al., 2020). Additionally, efforts are starting to shift to developing a micro-scaled generation of robots, one that could navigate high risk areas and continuously sanitize these high risk, high touch areas. With this development, sanitization could be more efficient, as the macro-scaled robots would be used to sanitize rooms and micro-scaled to target high touch areas. With the robotics' specialization, the macro-robots would need to spend less time cleaning each room.

These technologies can also be used beyond the healthcare

industry for infection prevention. Currently, there are many concerns about attending large conventions because there are limited means to ensure that the conventions are well-sanitized. However, with the utilization of such technologies, the likelihood of infections would dramatically decrease (University of San California Diego, 2020).

Robotic technologies currently used:

Many of these robotic technologies have been carefully tested and implemented around the world to reduce infection for frontline workers. Beyond private companies creating these technologies, many research institutes and universities have implemented these technologies in hospitals.

At the end of April, the Research Institute of the McGill Research Health Centre started to test an automated system to kill bacteria and other microorganisms, selecting the UVD robot to sanitize its equipment (Neustaeter, 2020). They started testing this robot on not only its effectiveness to clean hospital rooms and medical equipment and tools, but also stretchers and N-95 masks. The hope is to test if these N-95 masks can eventually be reused, rather than them being only once of use, providing a cost-economic solution to the lack of N-95 masks available globally (Neustaeter, 2020).

In May, an engineering company from Mangaluru, India, developed another UV-C light powered sanitization robot. This remote-controlled robot has already been deployed at Tejasvini Hospital in Mangaluru and is similarly effective as the other industrial robots, killing 99.99% of the microorganisms in approximately four minutes (Economic Times, 2020).

Keenon Robotics Co., one of the largest robot providers in China, has donated over fifty of its robots to areas throughout China and countries overseas in a bid to fight COVID-19. Its robot has similar lidar sensor and mapping technology that enables it to be smart enough to remember certain routes it has taken. In approximately fifteen minutes, this robot can remember its mapping route to effectively sanitize four hospital rooms repetitively. This robot stands out on the market because it also has the ability to return to its own charging unit, meaning further decreased infection to its frontline workers (Demaitre, 2020).

A Lithuania-based company, Rubedos, Slovakia-based company, Photoneo, and Singapore-based Digital Safety have all been working to develop disinfection robots, taking into consideration the expanding market for them due to the COVID-19 pandemic and future outbreaks of infectious diseases (Demaitre, 2020). Other universities such as Nanyang Technical University in Singapore, the University of Southern California in the United States, and Trinity College in Dublin have also been working towards producing disinfection robots (Demaitre, 2020).

Clearly, this robotic technology has become increasingly commercialized due to the COVID-19 pandemic and there has been emerging technology in this field all over the world. All these robotic technologies are more effective than traditional manual cleaning methods and they all fulfill specific niches in this market. It is vital in the next coming months that research centers and hospitals look at each technologies' purpose, benefits, and detriments to determine the best cleaning and sanitization robot for use given its purpose and goals.

Conclusion

With the plethora of innovative technology that has emerged in this particular sector in automated healthcare, it is safe to say that COVID-19 has pushed the world to the edges of innovation. Using technology to sanitize hospital rooms, tables, and other high risk, high touch surfaces, it has more effectively and efficiently cleaned these areas, killing up to 99.99% of bacteria for some of the technologies and can be cleaned in as fast as two to four minutes. Without a doubt, it provides further safety measures for frontline workers as sanitization personnel are not needed to directly contact these high risk, high touch areas. Additionally, since less personnel are needed for sanitization, they can be directed towards other areas of public care, meaning a greater focus on patient care.

However, it is also vital to analyze the extent that these technologies are effective and to determine its benefits and detriments. Despite its increased safety for frontline workers, it does not completely protect them and procedures must still be put into place to protect them and further mitigate the risk that presents frontline workers. Furthermore, since there has become a larger market for sanitization robots, it is important to recognize that each technology has its niche and it becomes essential that research institutes and hospitals recognize these niches, as it can help them sanitize their hospital areas and high risk surfaces more efficiently. With companies and research institutes developing this type of technology all over the world, this technology will most likely not only be used during the pandemic but for years to come. Thus, it is increasingly important that these measures and steps to vet each technology occurs because it will affect patient care and the staff's health long after this pandemic.

Robotics for the detection and sanitisation of viral hot spots

Robotics technology is not a new concept, as we have seen its applications to our daily lives, in different economic sectors, and COVID-19. We have already provided insight into its uses for diagnosing and preventing the respiratory virus, however we have yet to explore its uses for the detection and sanitisation of viral hot-spots. This is an important topic to consider because the diagnosis and prevention relies on the transmission of the virus, which mostly occurs at areas known as viral hot-spots.

The definition of these hot-spots vary greatly with the current research being done on the virus and therefore, they are dependent on the type in question. Nevertheless, the general consensus is that a viral hot-spot is a geographic area in which the prevalence, risk of transmission, and probability of disease emergence are elevated (Lessler et al., 2020). Some examples may include a particular city, a shop, or even a local tourist attraction. These areas are an obvious concern to the public, as the probability of people contracting the virus dramatically increases there. However, the problem with these spots is that they are only deemed as "viral hot-spots" once there has been a severe outbreak of the virus, which means that many people have already contracted the virus. Therefore, involving robotics technology in the detection of these hot-spots can be very beneficial to limit the spread and to ensure that the lowest number of cases arise from these locations. With the detection of these spots, the sanitisation is also an issue. This is because these areas must be sanitized very fast and effectively to control transmission. Hence, robotics technology will again be useful to perform these tasks. Robots would prevent frontline workers from sanitizing these locations, thus protecting them from the virus, and they also would not require the use of personal protective equipment (PPE) which is beneficial considering the global shortage.

The first aspect we are going to explore in relation to this topic is using robotics technology for the detection of COVID-19 in viral hot-spots. Now that we know about the importance of this for the safety of both the frontline workers and the public, we can look more into the progress that is done in integrating these technologies. During the earlier stages of the pandemic, the way in which COVID-19 hot-spots were detected was simply through testing. Then, they would look at the number of cases and trace the locations in which the majority of these infected individuals visited in the past two weeks. If there were a lot of cases arising from one particular area, then that area would be deemed as a hot-spot in order to prevent additional people from going there, reducing the spread of the virus.

Since then, many researchers have been trying to figure more effective ways of detecting COVID-19 in specific locations. One example is the proposal of implementing wastewater-based epidemiology (WBE). Researchers from Amazon State University (ASU) proposed this idea on a local and global scale. It involves analyzing sewage samples to determine information about human health (Chakraborty, 2020). This is a particular beneficial approach, as it helps to identify asymptomatic cases, providing a more accurate representation of the number of COVID-19 cases. By regularly using this method, scientists will be able to identify areas of outbreak much quicker, limiting the transmission of the virus to people. The testing of this approach has been underway and the results are good, indicating an optimistic future for this procedure. The main challenge in this is that workers would still be required to test the sewage waters, which would still require workforce personnel and PPE for detection.

However, there have been instances where robots have become integrated in this approach. In India, two social entrepreneurs developed a robot that can automate the process of tracking COVID-19 outbreaks through wastewater. Their robot is able to map and analyze various pipeline systems, which is the foundation of its processes. It collects wastewater samples, stores them, and then uses those samples to map out the areas of outbreak, all done while preserving the RNA in these samples. The robots are equipped with detachable modules to store the samples and are then sent to epidemiology laboratories to examine using RT-PCR analysis (Fluid Robotics, 2020). The entrepreneurs have partnered with Black & Veatch, a United States-based pioneer in the construction engineering industry. Their project is currently in the piloting phase and are projected to test out a prototype in Mumbai, India very soon.

This project has very obvious benefits. To begin, this model is extremely helpful for communities where there is a very dense population and have limited access to testing kits. This would mean that there is a very high population of asymptomatic individuals which would increase the transmission of the virus, putting more people at risk. Additionally, having robotic technology mapping out outbreaks is very efficient in terms of time, allowing outbreaks to be reported faster. Since this project is still in the prototype phase, the disadvantages are not yet known, however some common setbacks may be cost and the overall complexity of the system. The cost of implementing the robots can be too high, which may be a challenge for developing countries. Additionally, there is a degree of complexity of this approach, which may hinder its accuracy in terms of its performance. Since it will be going through wastewater, there is a risk that the robot may accidentally release the sample in open sewage areas. Open sewage water consists of untreated water and is a host to a variety of bacteria and viruses, and is already a problem faced by many countries, such as India (Chaturvedi et al., 2020). These spots pose a threat to people who come in contact with the water and therefore, it is important that the robots do not accidentally accumulate the collected samples in such areas. Nevertheless, the use of robotics technology in this way is a promising start to help detect COVID-19 cases in viral hot-spots.

Another project in the robotics technology industry that aims to help with the detection of viral hot-spots is developed from Miso Robotics, a United States-based technology company. They have designed a thermal-based screening device that is to be installed at restaurants. It attaches to the front doors and measures the body temperature of people entering the restaurant. It scans their faces and if the temperature is too high, the person would not be able to dine at the restaurant. This device is to be used for not only customers, but also delivery workers and staff of the restaurant to ensure effective COVID-19 transmission prevention. In addition to this, store owners can choose to receive text messages which would alert them if someone had a fever, for which they would deny entry into the restaurant. In this way, their technology can help prevent outbreaks of the virus at popular locations, preventing the formation of COVID-19 hot-spots (Wiggers, 2020).

As of now, this system is to be implemented at a popular restaurant called CaliBurger in Pasadena, California. If it is proven to be working as intended, Miso Robotics have claimed that they can install similar robotics technology in other public areas such as buildings, indoor tourist attractions, and offices. They also noted that data collection is anonymous, however if needed, the data can be analyzed by pooling it together to identify trends or patterns. This aspect of the system is particularly useful to determine the locations of COVID-19 hot-spots, as the data will show the areas in which the initial outbreaks of the virus occur.

Similar to the previous proposed project, this model also has its potential benefits and limitations. The main benefit is that areas where this project aims to be used include restaurants and other popular public places, which is important because it will help to prevent COVID-19 outbreaks. As a result, we may see an overall decrease in indoor viral hot-spots, which will prevent the spread of the virus. However, research shows that COVID-19 is transmissible even when the infected person is asymptomatic. Therefore, the extent to which this device may help reduce the number of hot-spots is still debatable, as it will not be able to detect the asymptomatic population. This includes people who do not have a fever but are still carriers for the virus. At that same time, having any form of virus detection method is still better than having no method at all.

Moving onto the robotics technology involved in the sanitisation of viral hot-spots, one example is from UVD Robots, a robotics company based in Denmark. They have sent robots to various hospitals around the world, including in China and Italy. These devices work by emitting ultraviolet light that decontaminates COVID-19 hot-spots such as hospital surfaces, as the light is able to tear apart the DNA strands of the virus. The robots do this by mapping out the environment to move around while being guided by lidar, which is UV-C light. The devices are able to kill 99.99% of bacteria and are also effective against viruses. They can operate for 2.5 hours on full battery, translating to about nine or ten hospital rooms. These robots are also being shipped to other countries around the world such as the United States and Canada, and they will be used to sanitise other hot-spots as well such as prisons, hotels, and airports (Robotopia, 2020).

The main advantage of these machines from UVD Robots is that they are effective and accurate. They do not require human interaction for operation and they prevent the need of staff to come in contact with the virus when sanitising and cleaning COVID-19 hot-spots. They are also very accurate, as they have been tested against similar viruses and are using powerful UV-C light, which is harmful to humans. Therefore, these robots are more accurate in the sanitisation of viral hot-spots, which make them useful in this process. In terms of potential disadvantages, one major drawback is that seeing the success and applicability of UVD Robots, many other companies have begun to release their own robots for sanitising viral hot-spots. The concern here is that there are many knock-off vendors in the market, which may not use the right safety precautions for operating the machinery (Demaitre, 2020). Robots like these will not perform to standard, which will put lives at risk. Additionally, these versions will damage the credibility of UVD Robots as well as the healthcare robotics community.

Other than sanitising surfaces, there are robots that are also performing the task of providing hand sanitizers and disinfectant wipes to people in public areas to prevent the transmission of the virus. Some even converse with people to educate them about the novel coronavirus and communicate ways in which they can remain protected. An example of this is the humanoid robot named Pepper, designed by Reliable Robotics from the United Arab Emirates. Pepper speaks fifteen languages and is able to interpret facial expressions, which as stated by the company, is perfect for raising awareness about the virus. He also dispenses a smart disinfectant solution in order to sanitize viral hot-spots, specifically in shopping centres, airports, or hospitals (Atalayar, 2020). Robots like Pepper are applicable almost anywhere and as a result, they can not only help sanitise viral hot-spots, but also work to prevent the formation of these through education.

To summarize the points explored in this chapter, we discussed what COVID-19 viral hot-spots mean in terms of its subjective definition as well as how we can identify them in our daily lives. Then, we looked at the need for appropriate detection and sanitisation methods for these hot-spots. For both of these aspects, the integration of robotics technology has proved to be extremely safe and effective in combating these issues. For viral hot-spots detection, robots that automate the process of tracking COVID-19 outbreaks through wastewater and that can help determine the body temperature of people have hopeful outcomes. In terms of sanitisation, robots that use ultraviolet light to disinfect hot-spot surfaces and those that dispense disinfectant solutions and hand sanitizers have also demonstrated their potential. Therefore, with the integration of robotics technology to fight the outbreaks stemming from viral hot-spots, we may be able to overcome this barrier against COVID-19.

Robotic Technologies for Nursing

Imagine what living in Canada would be like without our universal healthcare system. Our ability to be able to make in-person doctor visits, to purchase medicines from the pharmacy at low-cost, and to have long-term care provided to those who are disabled, injured, or in their old age, is a luxury. The 2020 COVID-19 pandemic has made us realize that access to free healthcare is something that we, as Canadians, have taken for granted.

Our healthcare needs have increased dramatically as we struggle to control the spread of the novel coronavirus in the country. We are in urgent need of support to nurse those who have contracted the virus back to good health.

You may have wondered: What exactly is the difference between a nurse and a doctor? How come our demand has increased for individuals in this professional field and not as much for those who hold a Doctor in Medicine during the pandemic?

Nurses are formally trained in providing care for the sick and have knowledge on how to diagnose, treat, and manage many conditions. Doctors on the other hand, are trained to treat diseases through medication, medical procedure, and in some cases, surgery (Jacksonville University, 2020). In light of the COVID-19 crisis, the need for nurses has increased. Their ability to provide care and innovation has saved lives and reduced suffering. With many patients currently in hospitals and critical care centres, nurses are at the frontline performing physical examinations, administering medications, and promoting good hygiene and social isolation practices. We commend our nurses for all of their efforts -- for performing above and beyond what is expected of them and for putting themselves at risk to secure the wellbeing of the public.

Unfortunately, our nurses are facing the worst side of the pandemic. They are always on their feet, working for a countless number of hours, which can have serious physical and mental implications. Their jobs are physically straining, with constant running back-and-forth to check on patients and to take care of their needs. By being in close contact with coronavirus patients, some nurses have also become victims to the virus themselves. In addition, they have been psychologically impacted by the number of fatalities seen. There is this constant pressure that nurses have to save as many lives as possible which can impact their ability to think critically in life or death situations. With WHO's announcement to the pandemic 'not even close to being over,' many nurses are frightened that our healthcare system may ultimately collapse.

Some of the nurses' fears have become a reality. Due to the skyrocketing number of cases, there is now a high patient - nurse ratio. In other words, Canada and many other nations alike, have a huge shortage of nurses. Our inability to provide sufficient care to coronavirus patients is leading to higher-than-expected mortality rates. Thus, new robotic innovations have been developed and are being implemented worldwide to help treat patients and reduce the burden on our human nurses.

The Use of Robots in Clinical Care:

With the global pandemic in effect, there is an urgent need for AI and robotic technologies. We have come to realize that humans cannot fight this pandemic alone. Our nurses are at the point of exhaustion and if we do not come up with viable solutions to help them, then we may see a continuation of increased deaths across the globe.

Robots have reportedly been used worldwide for COVID-19 in areas such as public safety, public works, non-clinical public health; clinical care; work, critical infrastructure, quality of life; laboratory and supply chain automation; and non-hospital care (Murphy, 2020). Some of the finest technologies have been developed for purposes such as healthcare worker telepresence, patient-doctor communication, disinfecting point of care, prescription/meal dispensing, and hospital administration.

1.Tommy The Robot Nurse

The first is Tommy, a robot that helps doctors in Circolo Hospital in Varese, Italy, care for COVID-19 patients. One of six robots, Tommy is placed in a patient's room to serve their needs, while doctors and nurses attend to others with more serious conditions. It has a monitor and a tablet with a microphone attached which allows for patients to be able to visually and auditorily communicate with healthcare providers remotely. This reduces the chances of the patient potentially infecting hospital staff who then have the potential to spread the infection at a much higher capacity due to the interactive nature of their work.

Tommy also has the ability to measure blood pressure and oxygen saturation for patients in the intensive care unit (ICU), including those who are attached to the ventilator system. Both measures are useful in determining a patient's current health condition and can be used to see whether the patient requires urgent medical attention from flesh-and-blood doctors. The development of these high-tech robots however, raises the concern that they may have the ability to replace most human workers, leading to grave economic repercussions down the line. According to Dr. Bonelli, director of Circolo Hospital, Tommy and other robots alike are "not a way to substitute the human factor, which is constantly something that [patients] rely on and we know how important it is for them." In other words, the robot nurses will not be replacing human nurses since human to human interaction is a key factor in patient recovery. Instead, they have and will continue to be used to help the hospital function efficiently with more incoming COVID-19 cases.

Apart from limiting the amount of direct contact that doctors and nurses have with their patients, Tommy and his high-tech teammates have also helped the hospital reduce the number of protective masks and gowns that staff have to use (Romero, 2020). This allows for the hospital to allocate its costs towards financing equipment such as ventilators and more robots that will be of urgent need to patients. The problem with implementing more robots in hospitals is the fact that their aim and function will have to be explained to staff and patients. For those that are not as tech-savvy, this can be a bit of a challenge, especially older patients who might not be accustomed to such technology. Providing an orientation on how to appropriately use the robots can take up additional time which the hospital may not have. Even though it may take some time to learn how to use and adjust to the technology, healthcare providers will not have to worry much about the robot's functionality. Robots, unlike humans, are not subject to exhaustion. Their batteries can be changed quickly after some time and then they are prepared to work once again. Their ability to serve for extended hours in hospital settings will be useful in providing consistent, high-quality, individualized care to COVID-19 patients.

2. Disinfecting Robots Used to Kill Coronavirus in Hospitals

Most individuals would avoid going to the hospital since there is a higher chance of contracting COVID-19 with the increased volume of patients. But this isn't really an option for healthcare workers and for patients already in medical facilities aiming to make a recovery. To prevent the spread of the virus in a hospital setting, it is mandatory for nurses to disinfect surfaces and all equipment. This can be a difficult and dangerous task, which calls for the need of autonomous robots.

A Danish company by the name of UVD Robots as mentioned earlier in Robotics for the detection and sanitisation of viral hot-spots, has made robots which are able to disinfect patient rooms and operation theatres in hospitals. They have the ability to disinfect about 99.99% of contaminants through their mobile array of short wavelength ultraviolet-C (UVC) light. It is said that the robots can "emit enough energy to literally shred the DNA or RNA of any microorganisms that have the misfortune of being exposed to them" (Ackerman, 2020).

UV disinfection technology has revolutionized cleaning methods in North America. It has been used to disinfect drinking water through a process called ultraviolet purification (Apec Water, 2017). Its success has inspired the invention of mobile disinfection systems that can be moved from place to place such as in airplanes. However for large environments such as hospitals, installing and operating UV systems can be difficult. It is not only expensive, but the machines are operated manually which can prevent some areas from being thoroughly cleaned. It is also important to keep in mind the numerous skin and eye injuries due to increased exposure of ultraviolet radiation (HealthLink BC, 2018). Thus, a disinfecting robot is a crucial discovery as it could be used to clear microorganisms such as the new coronavirus effectively and without human involvement.

These robots were under development for four years and were originally designed to address hospital acquired infections which are a problem internationally. According to UVD Robots CEO, Per Juul Neilsen, 5-10% of hospital patients will acquire a new infection while in the hospital, and tens of thousands of people die from these infections every year. The primary goal of these robots was to help hospitals prevent the risk of secondary infections. However, with the coronavirus in effect, spreading across nations, the need for these robots increased to fight off the pandemic in addition to other existing infections.

How has UV technology been integrated into the robots? The robot consists of a mobile base equipped with multiple lidar sensors and an array of UV lamps mounted on top. The robot is driven around the room with a computer, similar to the way that a remote control is used to drive around a car. The robot scans the environment using its lidars and creates a digital map. The hospital administrators can annotate the map to indicate the location of the rooms to indicate which ones have been cleaned and ones that have yet to be cleaned. After putting in these initial settings, the robot depends on simultaneous localization and mapping (SLAM) to navigate (Ackerman, 2020). It has a charging station which it will travel out of, and through the hospital on a daily basis, to perform disinfection tasks without human intervention. For safety reasons, the disinfection is usually performed early in the morning or late at night when people are not around. It has an automatic sensor to detect motion and will shut off the UV rays used to clean if a person enters the area.

It takes between 10 to 15 minutes to disinfect one room, with the robot spending a few minutes in a single position to maximize the exposure of UV rays on surfaces. In particular, the robot's UV system emits 20 Joules per square meter per second (at 1 metre distance) of 254-nanometer light which will practically destroy germs within a matter of a few minutes. The process used by the robots is faster and more effective when compared to human staff. It is also consistent since the robot follows the same path each time (due to digitized mapping system), which ensures order and cleanliness of facilities at all times.

However with such advancements to UV systems, there comes a cost. The robots are estimated to be approximately $80,000 to $90,000 US dollars (Ackerman, 2020). From a medical perspective, it seems to be a cost-effective technology due to the many benefits that it brings, including its contribution to reducing the risk of infection for patients and hospital staff. However, from a business perspective, it's important to consider the quantity that the hospital would need to clean the facility timely as well as initial set-up and maintenance costs.

We'd also have to be mindful that the robot is a machine after all and it may not have the intelligence to distinguish between microorganisms. This may pose as a problem in a hospital/healthcare facility where samples are constantly being collected and tested upon. If there is a way that the robot UV system can be digitized to allow for it to only disinfect potential microorganisms (in particular, COVID-19) and not any lab samples, then it may be a technology that may be of interest to a lot of facilities worldwide.

3. The Humanoid Robot, Cloud Ginger

With the spike in coronavirus cases, countries have been found to be understaffed and insufficiently equipped to fight off the growing number of cases. This has left medical professionals vulnerable to the virus which seems to have no cure. One solution, as with the rest of the technologies, is to reduce the contact between healthcare workers and patients who have contracted the virus. China was able to make this quickly happen with their invention of a "humanoid" robot called Cloud Ginger. The founding company, CloudMinds, had the approach of developing a robot that resembled a human so that an emotional connection could be created with it easily. By removing the coldness associated with the technology, hospitals have become more inclined to implementing Cloud Ginger as part of their nursing process.

Ginger has helped with hospital admissions, education services, as well in helping patients recover from the coronavirus due to its high-spirited nature (Clifford, 2020). In fact, it has been seen to dance and meditate along with patients to reduce anxiety levels. To accomplish its many functions in an engaging manner, CloudMinds integrated simple, child-like, and cartoon elements such as oversized eyes and a rounded smile onto the robot. This makes the visual and sensory technology also included in the robot, appear to be more user-friendly and personable. Furthermore, Cloud Ginger has 34 actuators that allow it to to behave in the ways humans move when they talk. This includes movements such as flexible fingers, arms that gesticulate in conversation, and the ability to make eye contact. Its voice has been designed to mimic human speech patterns through a thorough examination of pitch, tone, speaking cadence, and dialogue (Smith, 2020). Ginger's ability to read and understand human emotions, and to replicate them with a high degree of accuracy has allowed for the robot to empathize with patients, creating a trusting and loving bond with them.

Although the robot may seem to be a bit frightening at first due to its ability to provide care just as a human would, its use on a 24-hour basis can be helpful in reducing the burden on our hardworking healthcare workers. Its human-like appearance and nature can mimic the comfort provided by flesh-and-blood nurses in areas where there is insufficient care.

Final Thoughts:

Overall, nursing has been a crucial step in ensuring that coronavirus patients make a speedy recovery. However, considering how strenuous this profession can be, especially in light of this pandemic, the need for robotic technologies in our nursing procedures has increased. There have been countless technologies created and currently in the works, some of which include Tommy the Robot Nurse, which allows for telecommunication and performs patient checkups; disinfecting UV system robots and; Cloud Ginger which performs hospital duties and entertains patients. All of these technologies have benefited the way in which we have provided and will continue to provide care to patients. Although expensive, these technologies have streamlined procedures such as patient visits and check-ups, clean-up, and delivery, and have allowed them to occur in a safe manner.

Taking into account the success of these technologies during the pandemic, they may forever change the way in which we nurse patients for other current or future conditions. Automation, although a complex and time-consuming process, has led to valuable applications. If there's anything we've learned from the coronavirus, it's that we can't fight this alone. We will need more technologies such as robots to work alongside healthcare workers on the frontline for when this storm clears and for when the next pandemic arrives.

Robotic Technologies for Rapid Reconfiguration of Healthcare Manufacturing Lines

The pharmaceutical industry has played a huge role in fighting the COVID-19 pandemic. The production and distribution of live-saving medications and essential healthcare products have helped countless numbers of patients recover. Yet it seems that the coronavirus is causing more damage than expected, hard hitting our manufacturing sector. It all comes down to the fact that the way that our production lines function prior to the outbreak of COVID-19 is not sufficient to meet this increased global demand for healthcare items. Our medical supply chains have been overburdened with the pandemic, which has resulted in drug shortages worldwide. The national lockdowns to prevent the spread of the virus have delayed manufacturing and delivery procedures, making it difficult for patients to receive the products that they need timely (Banga, 2020). In short, there have been many fatalities as a result of our inability to cope with the pressures of the virus. All of these issues highlight the need for a more resilient global supply chain, one that can produce a range of items at a much higher capacity in a shorter period of time.

Robotic technologies can help change the way that our healthcare manufacturing system operates. In fact, some of the world's major suppliers of pharmaceutical ingredients including India, Switzerland, and Japan have already begun to pioneer technologies which may help address our growing healthcare needs amplified by the pandemic.

Cipla's Active Pharmaceutical Ingredients (API) manufacturing plant in Kurkumbh, India

The first of these technologies focuses on boosting productivity, which is crucial to ensure that drugs and healthcare products reach patients timely. Considering the stay-at-home orders for workers and restrictions on business operations, many companies have likely become a bit more laid-back and negligent in terms of ensuring product quality. Digital technologies in the form of advanced analytics, robotics, and automation can be used to reduce manual errors, increase consistency, and improve quality, which can deliver 30-40% increases in productivity (Banga, 2020). Cipla's Active Pharmaceutical Ingredients (API) manufacturing plant in Kurkumbh, India has exemplified the integration of such technologies in their supply chain.

Cipla is a company that primarily develops medicines to treat respiratory disorders, cardiovascular diseases, arthritis, diabetes, weight disorders, depression and other medical conditions (Cipla, 2019). With COVID-19 being their focus for 2020, the manufacturing plant has digitized production flow through a digital management information system. Specifically, Cipla has created a manufacturing execution system (MES) which is an information system that connects, monitors, and controls its complex manufacturing system and data flows. It automates material flow and captures cost information of products on the assembly line digitally in contrast to our traditional systems, which require the head assembly worker to keep a written record of the products being manufactured. By using a digital system to perform this process, it reduces any confusion and errors that could potentially be caused by a human worker, thereby ensuring the consistency of products. This system has also been programmed to alert workers of the status of the production line. It informs them whether they should continue to or halt manufacturing of a specific product, as well as if there are any issues concerning a specific product's quality. The function of this MES is to effectively execute manufacturing operations and to improve production output (Banga, 2020). It is important to consider however, that with such an advanced system, there are high maintenance and training costs required to operate.

In addition to installing the MES, Cipla has made changes to their drug assembly and delivering infrastructure. With the rigorous stay-at-home orders in India, many healthcare companies are beginning to develop coronavirus-detecting kits that can be used at the comfort of one's home. Cipla has robots which use force sensing to assemble all of the components of the medical testing kits which ensures that they are complete and safe to use. Robots are also placed at the syringe assembly lines to increase production of different sized syringes, all for the purpose of sampling and testing patients. It is a lot more cost-effective to use robots instead of purchasing new tools and employing specialized labour to produce different types of syringes. Lastly, robots are involved in packing and palletizing procedures in which they move quickly and can shift into different forms, which reduces delays, damage, or debilitating injuries that could result in human workers by working for hours on end in the manufacturing plant (Acieta, 2020). By employing robots, Cipla can reduce business costs associated with human labour while producing healthcare items safely and efficiently.

Noting its performance, US-based Gilead Sciences Inc. has signed a non-exclusive licensing agreement with Cipla for the manufacturing of the medicine Remdesivir, known for helping patients recover from COVID-19. Under this agreement, the manufacturing plant produces active pharmaceutical ingredients for Remdesivir as well as the finished product for distribution worldwide (Banga, 2020).

By developing digital systems which increase their efficiency and ultimately boost productivity as a large pharmaceutical supplier, Cipla can increase their competitiveness in the global API market. They can also use these technologies to mass-produce required healthcare products and medicines to satisfy national needs as opposed to depending on countries such as China or America for supply. This pharmaceutical supply chain will not only be able to extend aid to many coronavirus patients worldwide, but implementing these new technologies may be a unique selling point which will allow for them to generate revenue from both inside and outside the country.

Novartis partnership with Amazon Web Services

The second is not a specific technology being introduced, but rather, a partnership which has helped reconfigure the way in which Novartis, a Swiss multinational pharmaceutical company, is manufacturing and delivering its products to patients in need. Novartis has collaborated with Amazon Web Services (AWS) to address and deal with production challenges in light of the pandemic via cloud services.

Prior to this partnership, Novartis' manufacturing site metrics were centralized into a traditional Hadoop-based data platform which allowed for the storing of extensive operational data. Although data was periodically fed into the system and processed, operational reports were often generated based on these fixed datasets, which may not have been up-to-date with current operational decision-making (Partovi and Meyers, 2019). In other words, the system was not scaled to meet the operational needs of the business, and in fact, reduced the visibility that upper management had over the pharmaceutical company's manufacturing processes and distribution centres.

Novartis hopes to leverage AWS artificial intelligence (AI) and machine learning (ML) services to innovate different ways that drug and healthcare product lines can be reconfigured. In particular, the companies will work together to develop "Insight Centres" that will provide consistent and interactive operational information to both site coordinators and corporate users (Pharmaceutical Technology, 2019). Insight Centers display industrial sensor data which provides details on market competitor activity and success. They also contain AWS Internet of Things (IoT) monitored by computer vision algorithms, to provide data on the quantity and quality of materials and machines. The IoT services are specifically used for visual inspection of Novartis' pharmaceutical plants by capturing and storing images to track manufacturing risks. In essence, the cloud services provided by AWS AI can be used to capture inventory, quality, and production data.

The visibility of data that Insight Centers provide to Novartis will allow for the company to create predictive models for future production and to highlight possible challenges. The digitized system can recommend adjustments that can be made to the production lines on this basis to ensure the accuracy and quality of products. Each site Insight Center will also be integrated with the others, which will provide upper management with a global view of the production capacity (manufacturing status, inventory levels, costs) for the different healthcare agents/products used to treat COVID-19 (Partovi and Meyers, 2019). This information can be used to make informed decisions about production direction.

All together, the collaboration between Novartis and Amazon Web Services has facilitated informed decisions (through given real-time data) for efficient production and distribution of drugs and healthcare products. As Novartis' Chief Digital Officer Berrand Bodon says, "using data science and digital technologies to reimagine the way we manufacture medicines is not only at the heart of our transformation, but also core to our ambition to bring innovative medicines to patients faster" (Pharmaceutical Technology, 2019).

The following image summarizes how the partnership with Amazon Web Services will change the pharmaceutical supply chain at Novartis. By integrating the technologies that AWS has to offer, Novartis will be able to manufacture enough high-quality medicines to save more lives.

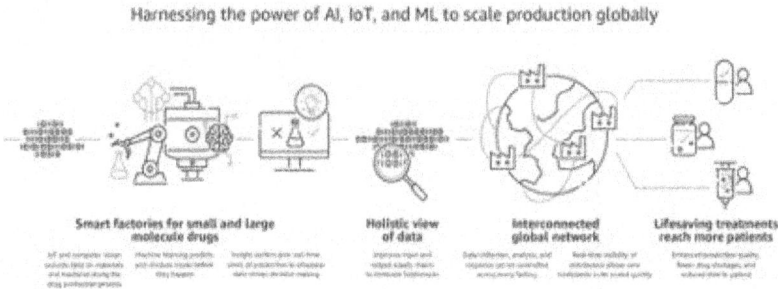

Harnessing the power of AI, IoT, and ML to scale production globally

Amazon Web Services 2020. Retrieved from: https://aws.amazon.com/blogs/industries/aws-and-novartis-re-inventing-pharma-manufacturing/

Musashi AI robots

With about 80% of the global workforce under social isolation, businesses, especially those that have turned to producing medical equipment, are finding ways to manage their supply chains to ensure stable and consistent manufacturing. The current crisis has resulted in manufacturing leaders such as Musashi AI taking a much closer look at how they can leverage technology to their advantage. They have designed robots with advanced deep-learning capabilities that have been trained to perform automated tasks within their factories. These tasks, which could take weeks or even months to train legacy systems to perform, can be learned within a matter of hours with new AI technologies. The first two robotic employees working in Musashi AI are a Visual Quality Control Inspector and a Fully Autonomous Forklift Driver.

The Visual Quality Control Inspector is able to identify defects in manufacturing lines quickly and accurately (Fretty, 2020). Musashi AI combined older ideas of computer vision with neural networks to teach the robot to identify faulty medicines. With cutting-edge optics and artificial intelligence, the robots can spot 99% of faults in less than two seconds (Fretty, 2020). The algorithms developed for this robot allow for efficient production operations and high product quality despite the restrictions on the business as a result of COVID-19.

The Fully Autonomous Forklift Driver robot navigates through the factory to perform material transport logistics tasks including collecting, loading, shipping, and receiving warehouse products. It is also responsible for identifying damages and reporting shortages or quality deficiencies with healthcare products (Fretty, 2020). The robot performs these tasks with high efficiency while adhering to safety standards to prevent injuries to any human staff that may be working on the factory floor.

Although these robots can make production operations more efficient, the initial costs and maintenance of new AI technology must be considered especially in the context of the pandemic, where many businesses are going bankrupt and cannot afford such technologies. We also need to ensure that these robots aren't taking over the roles that humans could perform. By redistributing much of the mechanical work to machines, the value of the human worker could be decreased, resulting in demotivation. This is ultimately the last thing that we would want along with worker shortages caused by COVID-19. We need to ensure that the AI being integrated into production lines is meant to reduce physical labour and keep workers safe. It must not replace humans, but rather, should prompt them to focus on other roles which will require their judgment and creativity, something which cannot be offered by machines.

Final Thoughts:

The pharmaceutical industry is one that has saved lives endlessly through all past pandemics. It will not fail us during COVID-19. There is no doubt that we are in a tough spot as we have seen production delays and drug shortages which have led to deaths that could have been avoided if we had addressed these issues earlier. To prevent further fatalities, we must develop robotic technologies to rapidly reconfigure our manufacturing lines so that we can produce high quality drugs and healthcare products timely. As said by Ran Poliakine, founder and chairman of SixAI, "the current crisis has been the catalyst for a change that has been on the doorstep of industry for a while now - a world in which AI can provide intelligent solutions for some of our most essential needs: food, energy, housing, and manufacturing" (Fretty, 2020). The 2020 pandemic is in fact, a catalyst for manufacturing businesses to become more efficient and productive. Incorporating AI services into production lines removes the need for humans to perform rigorous and repetitive work. The robots are flexible and consistent in manufacturing medical applications. With machines taking care of the manual labour, people will have the ability to showcase their true potential by engaging themselves in the development of processes and decision-making. We are in need of intelligent leaders that will make quick, but logical decisions which take advantage of the resources that we have at our disposal to deal with the pressures of the COVID-19 pandemic.

Cost & time efficiency of automation

As mentioned in the previous chapters, automation has become increasingly important during the COVID-19 pandemic, bringing up important concerns like safety in staying home versus going back to work to provide for family members. To effectively assess these situations, people need to weigh the related costs and time efficiency that comes with these automated processes against the benefits of going back to work and potentially risking infection. Some general pros regarding automation include: more efficient production, higher labour productivity, higher wages/profit, and improvements in safety and risk of human error. Additionally, manufacturing cheaper goods can increase the disposable income of consumers as well as allow for a shorter working week. On the other hand, some general cons of automation include: worker displacement and unemployment, potential increases in monopoly power, and loss of human interaction (Pettinger, 2019). Automation can also lead to more inequality, with industry executives benefiting from automation, whereas those in blue collar manufacturing suffer unemployment. With COVID-19, the use of automation has led to an increase in the number of tests per unit time, greater production in personal protective equipment for frontline workers, and other positive contributions to flattening the curve. This chapter will explore the cost and efficiency aspects of COVID-19 related automation processes, as opposed to the greater economic considerations.

Before diving into the COVID-19 related automation processes, it is necessary to look at the different types and aspects of automation to gain an understanding of the basics. Automation is fundamentally based on programming and relies on artificial programming interfaces (APIs), which allow communication between applications (Solis, 2014). If there is a web app in the browser, it will request for a search through the internet. This is where the API comes in, as it's the code that controls the various access points for the server, and allows for information to be sent from the web server database, through the internet, and back to the web app in the browser. The programming and APIs involved with automation allow for the integration of different systems together in order to act on scripted tasks. When combining automation with artificial intelligence (AI), machines can imitate human-like tasks including speech and pattern recognition, decision making, and visual perception (Eising, 2017).

The first type of automation is Robotic Process Automation (RPA) which primarily consists of software that automates manual tasks. RPA uses what are referred to as "bots" to automate mundane and repetitive tasks which allows businesses to allocate their efforts to more important work and reduce the time spent on costly manual processes. Similar to the paragraph above, RPA utilizes software to perform routine tasks across multiple applications and systems. These tasks can include the automation of transferring, editing, reporting, and saving data. Using RPA can improve accuracy, shorten processes from days to minutes, and increase efficiency for paper, manual tasks, and areas with complex workflow steps. RPA bots can directly collaborate with humans to achieve greater speed and accuracy in completing their tasks. These bots can also work independently and automate back-office activities that involve the collection, processing, and analysis of data (Tiernan & Ahmadinejad, 2020). With regards to COVID-19 and a hospital setting, this would be extremely beneficial, as a bot could collect data from hospital admissions, medical charts, and records to determine the amount of resources necessary at each location. Using this data, hospital administrations and resource providers could project how many ventilators and personal protective equipment are required at each facility and whether or not to allocate those materials to another, more affected location.

Another form of automation becoming increasingly prevalent during the coronavirus pandemic is hyperautomation. IT service management company Gartner describes hyperautomation as "an approach in which organizations rapidly identify and automate as many business processes as possible. It involves the use of a combination of technology tools, including but not limited to machine learning, packaged software and automation tools to deliver work." The company predicts that organizations will lower operational costs by approximately 30% by 2024. This would be achieved by coupling hyperautomation technologies with redesigned operational processes. Prior to COVID-19, Grand View Research estimated an RPA market size of $10.7 billion by 2027 (Solis, 2014). This could likely accelerate and change drastically due to the pandemic.

Surveillance and tracking citizens has become an effective way of limiting the spread of COVID-19 and incorporating automation into these methods would be an effective improvement. For example, during March in Singapore, the government was able to effectively contain the first wave of infections using surveillance cameras, police officers, and contact-tracing teams. These services have collectively aided the government in finding approximately 7,957 close contacts of confirmed cases, all of whom were quarantined (Palma, 2020). With regards to surveillance, over the past few years, artificial intelligence is being integrated into surveillance cameras, giving them "digital brains to match their eyes." This allows for analysis of live video with the presence of humans and can lead to more efficient surveillance. Unveiled in December of 2017, IC Realtime is an app and web platform named Ella that utilizes AI to create instant search results for what's happening in video feeds. It can recognize hundreds of thousands of language queries and allows users to search footage for clips of people wearing clothes of a specific colour, specific animals, or even particular car models and makes (Vincent, 2018). It is not difficult to imagine that these can be retrofitted to support searches for potential COVID-19 infected individuals, but these and other AI surveillance systems can be quite pricey, ranging anywhere from $100 to $2,500 depending on the company and mode (Amazon, 2020). What may mitigate the issue of cost, is the efficiency that comes with highly specific search results without the need and cost associated with human surveillance. Although with the Singapore situation, there were some concerns about privacy and invasiveness that were raised (Palma, 2020). This is a likely problem that will arise during these situations and in other areas that use this type of surveillance, prompting ethical guidelines and consensual agreement to be established.

Mobile phone apps have also become prominent during these few months in regards to tracking and collecting COVID-19 related data. For example, apps like AliPay and WeChat enable the tracking of citizens by the government and utilize a grading system ranging from green, yellow, or red health status (Grounds, 2020). As mentioned in Chapter 4, San Francisco health and technology company, Kinsa created a smart thermometer that pairs with a mobile app which retails for around $40 for the standard model and $70 for the ear model. Additionally, it has over 2 million users across the United States and approximately 150 thousand temperature readings per day (Bloudoff-Indelicato, 2020). These apps allow for extensive data collection in real time, but are limited in accuracy by what the user perceives and inputs as symptoms. Not only are they efficient, but these mobile apps are largely free and come with relatively inexpensive products, making them a good automation choice to invest in and further improve. Much like the surveillance systems, these mobile apps will receive pushback from individuals who would like privacy and don't want to give their information away. These problems will need to be addressed ethically and with competent guidelines as well.

Prior to COVID-19, the Shin-tomi nursing home and others like it in Tokyo, Japan were running trials for care robots in their facilities. Pepper, a robot developed in France, is used at the Shin-tomi nursing home and helps residents to exercise, monitor corridors during evening hours, and provide care and entertainment in the daytime. Similarly, Paro, a therapeutic robot seal, helps to comfort those with Alzheimer's in Japan and Tree, a motile, strong screen, helps those who are frail to walk. These robots and other initiatives present beneficial short-term implications during the COVID-19 pandemic such as mental stimulation for the elderly, but in the long-term, lack of human interaction could pose negative impacts on mental health. However, the UK Robotics and Autonomous Systems Network confidently projected a "further £6 billion productivity" gain for social care, but this comes with the cost the elderly, disabled, and those who experience mental health issues, as they will require more human interaction for verbal and emotional communication. The United Kingdom National Health Service (NHS) released a report estimating that automation could save 5.7 million hours of general practitioners time and save the NHS £13 billion per year (Grounds, 2020). This comes back to the argument of automation vs. humans losing their jobs which needs to be explored more deeply.

An area that currently requires a great deal of time efficiency is with testing kits and machines to quickly discern those who are infected with COVID-19 and allocate necessary resources to the most affected areas. As a recap of testing initiatives that have already been mentioned, there has been a lot of progress in Denmark, Germany, China, and parts of the U.S., with regards to increasing the efficiency and magnitude of testing per unit time. In Denmark, Flow Robotics created a pipette robot capable of performing 1000+ COVID-19 tests per day. Scientists were able to create a quick pipeline of twelve of these robots and use them in German and Danish hospitals to provide diagnosis for patients in just three weeks. These robots are very efficient and offer a user intuitive interface that allows anyone to learn how to effectively use them in less than an hour. Additionally, researchers from Renmin Hospital of Wuhan University, were able to create deep learning models that predict the likelihood of an individual contracting COVID-19 with 95% accuracy from CT scans of pneumonia. Although cases are incredibly high in the U.S., a popup laboratory at University of California Berkeley's Innovative Genomics Institute (IGI) created robotic pipelines to speed up the number of tests they perform. Partnering with the UC Berkeley Tang Center and medical centers around the Eastern Bay area, in a matter of weeks, they were able to set up a process that enabled from 1,000 to 3,000 tests a day. Results that would normally have taken from five to seven day to receive the results now only take 12 to 24 hours for the patients to know their results.

Moving onto robotic projects not mentioned in previous chapters, software and robotics company Bright Machines has been leading an initiative to develop a robotic system that can process COVID-19 test samples with little to no human presence. Automation in this scenario greatly increases the volume of tests done per day as robots can continually test for 24 hours, 7 days a week, as opposed to humans who would have to take breaks. Bright Machines has partnered with Tel Aviv's Ichilov Hospital Laboratory, Impact Lab and iCobots to develop the COVID-19 testing machine. This will use Bright Machine parts including a robotic cell, a dual conveyor, robotic arm, and a vision system. These parts in conjunction with one another, will allow for researchers to program the robot to open different-sized test tubes, drawing samples from one tube to the next. The vision system allows for verification of efficient performance in real time regarding the whole process. Green Machines and partners will provide the system to hospitals and facilities for no cost up to 12 months (Rayome, 2020).

Not only have robots been used to perform tests, but they alongside drones, have also been used to deliver materials ranging from test kits to groceries to products ordered from online. For example, at the Wuchang field hospital in China, front-line workers in a ward were replaced with 5G-powered robots to help mitigate the strain on human personal and aid with containment, preventing any spread of the virus. Additionally, companies like JD.com and others have increased production to get more delivery robots out into the open in order to deliver medical supplies within healthcare environments. These robots have also proved to be very beneficial in delivering essential items to people who are quarantined and cannot shop in person. Delivery apps like Meituan Dianping and SkipTheDishes have gained even more popularity than ever before, increasing their "contactless delivery" services through autonomous vehicles and robots. For those who choose to shop online, the companies that they buy from have also ramped up automation in their warehouses to streamline order fulfillment (Marr, 2020). Although the prices haven't been made apparent regarding these robots, drones, and other automation-related technologies, a relatively safe assumption can be made. Since consumer based models are around a few hundred dollars and the expectation is that the models for hospital use (robots) or delivery (drones) will need to be more compact, faster, and efficient, the costs for these products may range anywhere from $1,000 to $100,000 (Amazon, 2020).

With the healthcare systems of many countries being overwhelmed by the pandemic and frontline workers being swamped with patients, it is important to help keep these heroes as safe as possible by implementing automation into the production of personal protective equipment (PPE). During May of 2020, an Ontario-based manufacturer of fibre-reinforced plastic products was able to develop an effective COVID-19 testing station to reduce the use of PPE by the general public. The station is called the IsoBooth and is very efficient when it comes to making the most out of the limited PPE. The healthcare worker or testing administrator is stationed in a booth that is made of fibreglass and polycarbonate and has two holes that lead into arm length gloves on the outside of the booth. This eliminates the need to dispose of masks, gloves, gowns, and face shields after every test is administered making it highly efficient. A quote from a manufacturing article states "Ontario alone intends to ramp up testing to 16,000 per day. Implementation of the IsoBooth would eliminate the use of almost two million disposable gloves and one million masks in two months." Although testing does require a human, these booths are able to significantly reduce PPE wastage and by automating the production of these booths, the benefits significantly outweigh the costs. For example, it is estimated that approximately 3,900 kilograms of PPE hazardous PPE waste is generated everyday from hospitals and other healthcare facilities across Canada for every 60,000 tests administered nationally (Manufacturing Automation, 2020). The IsoBooth helps to significantly reduce the magnitude of pollution, our environmental footprint, and the stress on waste treatment management facilities as well.

The province saw even more positive contributions during the next month of June, when an Ontario PPE manufacturer began to automate surgical mask production. The Canada Shield, a Waterloo, Ontario-based PE manufacturer started producing not only reusable face shields and cloth masks, but also disposable surgical masks via automation. This will allow masks to be produced 24 hours, 7 days a week to ensure that the company meets the production quota of 1 million surgical masks per week. The Canadian Shield's annual capacity is expected to grow to more than 200 million surgical masks annually in the next two months (Manufacturing Automation, 2020).

Automation has seen an unprecedented acceleration with regards to implementation in a myriad of industries due to the COVID-19 pandemic. The most applicable types of automation to healthcare related innovations are Robotic Process Automation (RPA) and hyperautomation. RPA's are used to perform repetitive and mind-numbing tasks whereas, hyperautomation refers to automating as many business processes as possible in order to achieve maximum efficiency and productivity. Automation has been coupled with AI in surveillance to effectively track and prevent the spread of the virus as well as aiding in creating disease models and compiling symptom data in mobile apps. Many robotic pipettes, booths, drones, and other testing implementations have been created to increase the pace of testing and deliver essential PPE to healthcare providers without risking the spread of COVID-19. Finally, many manufacturing companies, especially in Ontario, have ramped up PPE production via automation to mitigate the strain felt by healthcare systems across the globe. Many of these innovations have led to monumental increases in efficiency, but can be costly to manufacture. From this point, those in charge must weigh the benefits of increased efficiency against the costs for each automated-related solution to determine what is the best course of action that will lead to positive change.

Economic considerations of automation

We undoubtedly know that automation is the future and it will be integrated in our daily lives in many ways. This is because the passionate and advanced ideas that we have as a society can only become a reality through the use of technology. Our drive to make lives easier and tasks more effective is what will fuel the everyday use of automation. In general, automation has many benefits. Most of these advantages are those that we are familiar with already. However, like most changes, there are always aspects in which you win some and you lose some. The use of automation is no exception. Although there are many obvious benefits, there are important limitations that need to be considered and recognized to ensure that we use automation of its full potential. In this chapter, we will specifically look at the economic considerations of automation with perspectives on the employment sector, the growth of the economy, the income and wealth inequality, and the implications on the global economic scale.

To being with the employment sector, there are many pros and cons when it comes to automation in a workplace environment. The first advantage is overall improved productivity and growth, as tasks that humans are being employed to do can be done at a faster and more efficient rate through the use of automation. Workers are humans too and they need breaks and have a certain limit as to how fast they can do tasks. However, robots and other robotics technology can perform repetitive tasks continuously, which is very beneficial for businesses and workplaces, as they not only can save their costs of labour, but also receive a better labour output thereby generating more profit. In terms of the healthcare industry, robots can overtake jobs such as cleaning and administration tasks, which would prevent labour. Next, even though the everyday tasks that humans perform can be automated, there are tasks that pose a risk to humans. In this case, having robots or automated machinery carry out these duties results in a safer workplace environment for employees. Such tasks include those that involve working with dangerous chemical exposures, welding, and blades. To add onto this, when humans working these jobs are required to work for long hours, this increases their chances for mishandling errors that may result in serious injuries. Therefore, the automation of these tasks results in more accurate performance, as there is less room for these errors. An example in healthcare is the exhaustion of frontline workers during COVID-19. They have to battle with the virus along with taking care of patients and performing necessary duties that have the potential to be automated. Finally, another benefit is the increase in employee value. When basic and dull tasks are automated, this leaves room for more creative employment opportunities. This would lead to more enjoyable jobs, which would likely result in a higher worker morale and better employee output as well. In terms of the current pandemic, if some tasks of frontline workers become automated, then that would allow workers to focus their attention on other areas where COVID-19 has had a negative impact, such as nursing homes and hospitals.

On the other hand, there are significant cons to consider when looking at the impact of automation on employment. The first is employee displacement, which occurs when automation replaces the jobs of certain workers. This results in unemployed workers, which means that they have to find another job themselves or be displaced to other roles by the employers. Understandably, employee displacement has the potential to cause great emotional stress, as finding a new job and getting adjusted to a new routine can become challenging. Therefore, in order to implement automation, employers must be cognisant of this issue and must make the transition for their employees as smooth as possible. Furthermore, when automated systems begin taking over certain jobs, the employment opportunities will also change. On one hand, it may create more innovative and thought-provoking opportunities. While this seems more enjoyable and interesting, it will require a higher degree of education or experience, which may result in those without these qualifications to struggle to find appropriate jobs. On the other hand, employment opportunities may include more jobs with roles to operate, manage, or overlook these newly implemented automated systems. Additionally, the jobs that would supervise the machinery have their own challenges. People employed in these positions may feel as though they are becoming "slaves" to machinery and technology. This can take a toll on their mental and physical health, as it can demoralize their sense of self. Therefore, employers must also ensure to focus on their workplace personnel to ensure that they remain motivated and valued, regardless of their positions.

An impact automation may have is the increase in economic growth. Economic growth is defined as the increase in the production of goods or services of a business overtime. This growth may occur as a result of increases in capital goods, workforce, and technology. It can be measured by various mathematical estimates, such as gross domestic product (GDP) (Chappelow, 2020). In terms of the effect of healthcare automation, the economy will grow, as there would be an increase in the workforce and technology. The introduction of technology will boost productivity, as regular tasks will be done by the new systems and the labour force can then shift their focus on ensuring that patients receive the best care. They would also be able to provide additional services which may benefit the experience of receiving treatments, further boosting the healthcare economy. At the same time, even though there would be an increase in workforce, it is important to note that this increase will not occur overnight. In fact, many business-focused sources have warned that the initial stages of implementation of automation will disrupt local economies (Aspen Institute, 2019). It is highly likely that the overall workforce will decrease initially, as the current labour force becomes accustomed to their new roles. Hence, it is important to consider that these changes would take time and it is a long-term progress for the economy, rather than an instant growth.

Another way in which healthcare automation would result in the growth of the economy is the creation of jobs. This topic was touched upon in the beginning of this chapter, however, the other method in which jobs can be created is due to the increase in goods and services. It is a fairly simple idea to consider -- it is like a cycle in which the end result is more opportunities for employment. However, when economic growth is discussed through perspectives like this, we often do not consider those factors that are in fact not measurable by GDP. One such example is reluctance to automated interactions that some consumers may face. The loss of the "human touch" may be enough of a reason to limit consulting healthcare services, as some may not believe that robotics technology has more knowledge and expertise than a certified health professional, thereby affecting the economy negatively. Therefore, even though there may be strong and alluring economic benefits, we must consider the degree of automation in different industries. Some fields, such as retail, may be more beneficial to implement a higher degree of automation than others, such as healthcare. For this reason, it is helpful to have an open conversation about the extent to which automation should be implemented, one that brings in the perspectives of those working in the field, consumers, and the general public. Therefore, in order to continue economic growth, it is important to consider factors that cannot be measured as well.

Moving on, another economic aspect to consider about the implementation of automation is the rising income inequality among the labour force. This is because new technology will have different impacts on jobs. For high-skilled workers, these new technological systems will have a more productive impact when compared to the low-skilled workers. This concept is called Skill-Biased Technological Change (SBTC) and 85% of economists believe that SBTC will be a major cause of inequality in the near future (Admin, 2020). Therefore, the incomes of the high-skilled workers, which typically have a higher education and more experience will increase, while the incomes of the low-skilled workers will remain stagnant. This trend has already been seen with the job markets in the United States of America, which shows that there has been a 70% raise in income for males with Masters' and Doctorate degrees, while the wages of male high school dropouts haven't significantly increased (Admin, 2020).

One a similar note, if income inequality increases, then it is projected that the average household wealth inequality will increase as well. This is because automation would result in products and services to be cheaper, hence the economic gains can then be distributed in three ways. One is that there would be less prices on products due to economies of scale, thus benefiting consumers. Next, it can benefit workers through higher wages, and finally, it can increase the profit margins for the owners of the business. Different businesses would distribute the gains in various orders, which would ultimately result in the inequality of wealth. Additionally, another phenomenon known as "deepening automation," can also contribute to this inequality. Deepening automation is used to describe when a task that is already automated is made even more productive through technology. This has the potential to then displace even more of the labour force, which would result in lower economic benefits for these workers overtime.

A final economic consideration to explore is the integration of the world's economies. This concept can be described as globalization, which is the increasing interdependence of the economies around the world. This occurs when businesses begin to operate on an international scale and therefore have a form of global influence (Kopp, 2020). Due to these difference economies becoming incorporated with one another, it is usually seen that when one economy has a downfall, it creates a domino effect that ends up affecting its trading partners from different economies. Similarly, if one thrives, it can have a positive influence on those economies that are interdependent. Therefore, any effect on one country's economy is bound to have an effect on other countries' economies. In this way, the impact that automation will have on the economy will be translated on a global scale.

In conclusion, there are many economic factors that must be considered during the process of implementing automation in our industries. The main factors revolve around employment and the job market. We explored the main pros and cons involved with this perspective, with the main pros being improved productivity and growth, preventing harmful risks to workers on the job, and an increase in employee value. The main cons include employment displacement, change in employment opportunities, and the emotional challenges involved with the integration of automation in the workplace. Next, another economic consideration is the growth that would occur and how this growth comes about. This includes looking at how automation would lead to an overall increase in high-skilled jobs, as more education and experience is likely needed for employment, as the low-skilled jobs are being targeted by robotics machinery. Looking more into this specific topic, we explored how this imbalance between high and low-skilled jobs can lead to inequality in the job market. This inequality can make it harder for displaced workers to find jobs and for those with limited access to educational opportunities. This income inequality also translates to wealth inequality, as the average wealth in a household will also be different, as those with access to education and certain opportunities would have a higher overall wealth compared to those who do not. All in all, to ensure that the implementation of automation is as successful as can be, these factors must be taken into consideration, especially considering the concept of globalization. Specifically, plans for the displaced labour force must be put in place to ensure that current and future employees are not placed at a disadvantage and can still equally contribute to the economy as their counterparts.

Psychological effects of automation   lack of human interaction

Humans, by nature, are social creatures. Going back approximately 200,000 years to the dawn of our species, tribes began communicating with one another to form larger social groups, hunt, and provide resources for one another, enabling them to have greater chances at survival. From here, Darwin's idea of natural selection took over and human communication was subjected to thousands of years of compounded evolution. In the present day, humans require social interaction, as it is necessary for their health and mental well being. Newborn children who are given all the necessary elements to survive (food, water, shelter), but love, will either die or have a very high likelihood of developing long-term psychological damage (Szalavitz, 2010). The need for consistent social interaction has become increasingly prevalent during the COVID-19 pandemic, as many countries have issued their citizens to quarantine for several months to reduce the risk of spreading the virus. Those who actually become infected may have to quarantine even more strictly (not be able to see anyone) for longer periods of time. The uncertainty of when one will be able to see and hang out with their friends and loved ones has had very important psychological implications for all age groups.

Additionally, with the heightened rate of automation, many people have lost their jobs either indefinitely or for good, which can have huge impacts on human health and general well-being. A 2013 Oxford Study estimated that almost 50% of jobs could be at risk of automation by 2033 and that changes in health could increase costs for health providers by millions of dollars, including up to $47M due to stress related to job-security alone (Riley, 2018). It is also important to note that although less than 5% of occupations are 100% automable, 30% of the work involved in most jobs can be performed by machines (Hewitt, 2017). There is a clear relationship between job status and mental health, as 18% of American adults are treated for depression after remaining unemployed for 27 weeks or more (Richardson, 2017). This chapter will explore the different psychological effects of automation, mention more recent figures, and discuss the importance of examining impacts on human health before quickly rushing to automate thousands of jobs.

It is important to note the bias against the costly and tiring nature of effort displayed by cognitive psychology, neuroscience and economics. This bias may actually downplay the risks that boredom brings as well as underestimating the value of exertion. "Effort can be defined as the subjective intensification of activity - mental or physical - in the service of meeting our goals." In trying to develop ways to reduce human effort, people often overlook its benefits. A common example comes through The "IKEA effect", a cognitive bias in which people will be prepared to pay more for objects that they have partially created, as opposed to objects that they didn't have a role in creating. This brings up the effort paradox, where effort can be valuable and rewarding in its own right, offering intrinsic motivation. The effort paradox and bias of effort become important in regards to automation, where boredom may become an increasingly pressing issue. During the beginnings of the Industrial Revolution, German philosopher Friedrich Nietzsche voiced concerns regarding a "machine culture" which would create extremely mundane conditions for workers. A study from 1980 described boredom as "halfway between misery and sleepiness", characterized by low arousal and dissatisfaction. Many industries and businesses may have to reconsider automation, as it could have detrimental impacts on people's well-being and performance. A 2017 study used electroencephalography (EEG) to monitor the effort and boredom on patients' brains. Results suggested that boring tasks can actually be more fatiguing with increasing time, as compared to continuously exerting cognitive effort (Hewitt, 2017). Although humans may not be displaced from their jobs, some degree of automation will be present that could create an increasingly dangerous environment of boredom.

As one delves more into the psychology entangled with automation, some interesting ideas and thoughts arise. A study published in the academic journal Nature Human Behaviour, looked at the effect of automation on people's attitudes to being replaced. Conducted in 2019, 2,000 participants across North America and Europe were asked to imagine a variety of different scenarios which ranged according to whether it was the participant's own job or the jobs of others that were under the risk of being automated. Interestingly, the study found that "while people tend to prefer the jobs of others to be replaced by humans, when it came to their own jobs, people would rather be replaced by machines than another human" (Makhzani, 2020). Having looked at dated demographics in the introduction of the chapter, it's important to look at the more recent ones. The World Economic Forum predicts that, with the rise of artificial intelligence, some 375 million jobs will be lost, as 60 to 90% of all jobs now in place will be impacted by artificial intelligence (Manyika et al., 2019). 50% of human beings across the globe are symptomatic of mental illness at some point in their lives - 20% in any given year. AI coupled with the atmosphere generated by COVID-19 could spark an array of consequences and turbulent times, causing those depressive symptoms to arise. However, AI has promising applications in creating technologies to diagnose and treat disorders like schizophrenia more effectively. In 2019, the Government of Canada invested heavily in AI systems to enhance internal decision making by introducing a Directive on Automated Decision-Making and an "algorithmic assessment" to sort out the ethical responsibilities for using AI (Wilkerson, 2019).

The impact of automation varies from country to country across the globe. It is generally thought that the Global South is more severely impacted by automation than the Global North. This is because the more rich and developed countries have already outsourced the most easily automatable tasks to the Global South, where manufacturing jobs were growing until recent events. Companies are reshoring their production facilities to save money by exploiting the situation of rising labour costs in the Global South and advancing robotic technologies in the West. This allows machines to return to their mother country and do most of their jobs from there, which is referred to as lights-out manufacturing as absolutely no human presence is necessary on site. With supply chains being strained due to the coronavirus pandemic, reshoring operations could potentially be accelerated in an attempt to meet product quota and related demands (Witte, 2019). The nature of international economics also becomes influenced by all the automation efforts, as countries will need to work together to benefit each other's supply chains and be able to provide for their citizens without creating any conflict.

When looking at different age groups and the impact of automation, it seems that the elderly are among the most affected. Specifically, those situated nursing homes, as they normally don't see much social interaction apart from the health care workers, others in the facility, and occasional visits from family members. Due to COVID-19, this social interaction diminishes to almost zero, as the elderly tend to have weaker immune systems and are at a much greater risk from the virus. Even prior to the pandemic, as pointed out by Professor Belinda Bennet, "changing demographics have led to increasingly challenging programs in elderly care, especially those with dementia." It is possible however, to meet these challenges by increasing contact with family members, carers, and the community. The elderly care industry in North America already sees some preexisting conditions that have limited social interaction. Due to the nature of the laws regulating filial piety, many have dropped off their parents suffering from dementia to nursing homes without their full consent, or even lied to them about regular visits. The implementation of technology that further relegates human tasks will contribute to the continual removal of a human element and connection from these patients who so desperately need it. However, in Asia, countries like China have established an "Elderly Rights Law" requiring children to visit their parents, but with the COVID-19 situation, visits can still happen via technology instead of in-person. For example, using realistic virtual reality and AI managed chat-bot systems based on the conversational habits of a real person could greatly help to maintain social interaction (Dong , 2017).

Shifting back to the working population, but focusing on the United States, this study looks at the evidence of county-level job automation risk and health. Conducted in 2018, the study remarks that workers in the United States have increasingly become more uncertain about their future employment in recent decades. Similar to the estimates in the introduction, this study puts the high estimate of potential job losses due to automation at around 47%, signalling that job security risk due to automation is significant and growing. The study uses a preliminary model to test whether high automation risk leads to poor health outcomes with perceived job insecurity as a mediator. (automation risk - job insecurity - poorer health) Since automation fuels fear and anxiety of job loss, then it may compound anxiety over job security. After reviewing 57 longitudinal studies on job insecurity and health/well being, De Witte et. al (2016) concluded that "job insecurity affects health and well-being on the long-term rather than the other way round." Other studies have found that anxiety coupled with other aspects of job loss significantly increases negative mental health indicators, including exhaustion/burnout and depression. The overall study found a negative relationship between county-level automation risk and county-level general, physical, and mental health (Patel, Devaraj, Hicks, & Wornell, 2018).

Although briefly mentioned in the previous paragraphs, automation has recently been used to detect mental illness with artificial intelligence. In 2019, Researchers from the University of Colorado Boulder were working to apply machine learning artificial intelligence in psychiatry via a speech-based mobile app. Through a number of tests, the app has successfully been able to categorize a patient's mental health status just as well as, or even better, than a human can. Specifically, patients are asked to answer a 5 to 10 minute series of questions by talking into their phone. They're asked a variety of things ranging from emotional state Peter Foltz, research professor at the Institute of Cognitive Science, co-authored a paper about AI in psychiatry in the academic journal Schizophrenia Bulletin, and stated "We are not in any way trying to replace clinicians, but we do believe we can create tools that will allow them to better monitor their patients." The World Health Organization (WHO) estimated that 44.3 million people suffer with depression and 37.3 million suffer with anxiety in Europe alone. Unlike a blood test, diagnosing a mental health disorder is based on an age-old method that can be subjective and unreliable. Brita Elvevåg, a cognitive neuroscientist at the University of Tromsø, Norway says "Humans are not perfect." There are many windows to make errors and miss out on subtle speech cues and warning signs. Elvevåg and Foltz worked together to develop machine learning technology that can precisely detect day-to-day changes in speech that clue at mental health decline (Price, 2019). However, problems around public trust and proving efficacy arise that need to be addressed prior to mass implementation.

Discussing the psychological effects of automation brings up a wide array of important and related topics. The effort paradox of enjoying something more when exerting more effort as well as the idea that boredom is halfway in between sleepiness and misery, brings up important concerns about human health and job satisfaction. The increasing prevalence of automation also sheds light on human nature and the fact that people would rather have their job replaced by a machine instead of another person. Not all areas are similarly affected by automation. The developed countries of the Global North have previously automated their facilities in countries from the Global South, causing the workers in these facilities to lose their jobs. Because of the coronavirus pandemic, many of these facilities have been reshored in order to save costs and mitigate the strain on supply chains. Countries from both the Global North and South will need to cooperate amidst the accelerated automated process to maintain economic relationships. Additionally, the elderly population are the most at risk demographic and thus, are experiencing more social isolation and detachment from human elements. Automation is accelerating this process of detachment and further limiting the human interaction the elderly so desperately need, especially those with neurological disorders and conditions. Nonetheless, there is still some silver lining associated with automation that comes from the field of artificial intelligence psychiatry. University researchers across the globe have been implementing machine learning artificial intelligence in diagnosing and monitoring the presence of mental disorders in patients via mobile apps and computer generated algorithms. With all the negative psychological effects associated with automation, there are still promising technologies that contribute to bettering human health. Governments and businesses will need to ethically weigh the costs and benefits of automation before implementing it in their processes, assuming a necessary human health perspective

Privacy & security concerns
Confidentiality & ethics

Is my location being shared? Is my smartphone secretly recording my conversations? Do I know if my financial data is being compromised? Questions like these are often wondered and even asked, yet sometimes the answers are not so simple. With most of our lives being online, there are many valid privacy and security concerns. However, with the increase in artificial intelligence, automated systems, and telehealth, these concerns are more important to address than ever. This is important to consider because those behind this technology have to ensure the full security of their consumers. Automation in the healthcare industry would include the collection and storage of highly personal and sensitive data. This in itself, makes it more crucial than ever to have dialogue about confidentiality involved with the implementation of technology in healthcare.

There are many examples that highlight this importance. One such example is the WannaCry cyberattack. WannaCry is a computer application that was able to break into the United Kingdom's National Health Service (NHS), as a ransomware target. The hackers were able to make use of a flaw in Windows, an operating service, to breach into sixteen health centres and about 200,000 computers. This attack resulted in the cancellation of around 20,000 appointments and created damage in 1,200 diagnostic equipments (Genovese, 2019). The overall effects of this breach were devastating, as it resulted in delayed appointments for patients with serious needs and incurred a lot of time and costs for the United Kingdom's healthcare system.

The effects of cyberattacks on hospitals and health services are clearly immense. However, the compromisation of patient records is a significant issue on its own. In 2019, the total number of exposed patient records was 41,335,889 in America, compared to the 13,947,909 records that were compromised in 2018. To estimate an idea of this astonishing growth of breaches, the report details that "more healthcare records were breached in 2019 than in six years from 2009 to 2014" (Writer, 2020). One of the biggest healthcare breaches that occurred in 2019 was the American Medical Collection Agency (AMCA) data breach that resulted in the release of records of 25 million patients. AMCA is a billing services vendor and was hacked for eight months, from the beginning of August 2018 to the end of March 2019. The data that AMCA stores and that was breached included financial information, such as credit card numbers and bank account information, Social Security numbers, and private medical information (Davis, 2019). There were many companies with these records that were hacked, which resulted in this staggering number of accessed patient records.

Other than cyberattacks and software viruses, there are many other factors that contribute to keeping patients' personal and medical information private and confidential. Most of these factors revolve around the idea of online transmission. This includes having audio or video calls with a healthcare professional or sending medical and financial information online. There are many risks of such information leaking by being easily hacked through signal interference or other transmission interruptions. Many people would be rightfully concerned about this because these events are not as uncommon as we would initially think based on the numbers already mentioned. Therefore, healthcare automated systems would require very strong connections and encryption in order to ensure for maximum security. In fact, it is recommended that the transmissions are all encrypted and that the connections have two-factor authentication, rather than one-factor. Additionally, patient interactions with automated healthcare systems should occur in private virtual rooms in order to secure privacy and confidentiality of the patient's information (McGee et al.).

To add onto the stress about keeping patient confidentiality, automated healthcare offers more challenges. The main one is that the patients, healthcare professionals, and the general public will not know exactly how their data is stored. We know that data would be stored in databases and through secure software, however, when patients feed their information to a technological device, they are not immediately aware of this. There will always be a sense of curiosity in this case compared to when we are in a doctor's room. There, we know that we are physically telling our doctor information and that they will be keeping this information confidential. However, with a computer, we do not know for certain where our information travels and how our information is analyzed. This can be a sense of worry to the patients. For example, they may think that they might have inputted information wrong or maybe they gave too much information, or too little. Therefore, it is important that the automated programs that we develop have very clear instructions to the patients, which would detail exactly how much information to give, where this information is being stored, and how it will be accessed and subsequently analyzed.

Even though the healthcare automation industry must do everything from their end to address the privacy and safety concerns of their consumers, their patients should also be doing the same. The responsibility on them is that they must ensure their interactions with such automated systems are in a private area, and must be done in the presence of nobody else. Additionally, the patient must ensure that their computer or technological device is secure. This would involve having a strong connection to a safe network, having an updated virus protection program, and ensuring that the websites visited are official and pose no threat to the computer. Just like the automated devices, patients must make certain that they also have encryption and complex passwords to have controlled access. Patients surprisingly have just as a role in maintaining the privacy and confidentiality of their information as much as the automated healthcare industry. Therefore, the patients must be aware of the potential vulnerabilities of their home devices and try to limit them as much as they can.

Another challenge that comes with the privacy and security concerns of automated healthcare is fraud. There may be third parties which will charge patients for prescriptions and consultations that never occurred, and thereby performing fraud to make financial gain. This occurrence has specifically been seen during the COVID-19 pandemic. With the expansion of online and automated healthcare services, fraud events have shot up dramatically during this time. One particular example is the fake COVID-19 tests that are being sold online (U.S. Department of Health and Human Services, 2020). What seems as a legitimate healthcare facility, these fraudsters perform identity theft to collect personal and financial information from people in exchange for the unapproved tests. This simply goes to show how frequently these identity thefts actually occur. When healthcare services become more automated, it is reasonable to assume that these identity thefts would increase and the fraud rates would follow suit. Therefore, it is important that fraudsters are monitored and removed as soon as possible, and that the public becomes aware of the difference between official healthcare organizations and fake facilities.

Furthermore, the next potential risk in the automation of healthcare is the risk of misuse of data. We have seen this happen with Facebook with the collection of data. Allegedly, Facebook was collecting information including gender, birthdates, and other personal information to personalize advertisements for its users. This event raised privacy concerns, as this was done without the consent of its users. We have also seen this occur with health information through genetics testing companies, such as 23andMe. Genetics testing companies work by analyzing DNA to then provide its users with personal information such as their health, traits, and ancestry (Lexalytics, 2019). Although consent is given in this situation to allow them to store your DNA for up to ten years, regardless, there are still some events happening without informed consent. One of these is that insurance companies can be using the genetics data to bias their pricing and selection. This is because this genetics test also determines health risks of their users based on their genetic predisposition. Thus, experts warn that insurance companies are using this data to set higher prices and have more limited selection for financial gain. All in all, it is important that people are informed as to what is going to be done with their information. This goes back to the importance of informed consent, which includes users to be fully aware and comprehensive of the storage, purpose, and access of their data in order to prevent the misuse of their personal data in the future. The next issue that has been deemed as a threat to the implementation of automated systems in healthcare revolves around the idea of transparency. It is said that many artificial intelligence algorithms, specifically though used for image analysis, are unfeasible to explain. This means that if a robot analyzes a picture of discoloured skin, for example, and then provides a diagnosis that the patient has skin cancer, there is no way to determine what led to that diagnosis (Davenport et al., 2019). This raises an ethical issue, as we need to determine how much of ourselves are we willing to feed into these machines in order to receive a diagnosis that has no explanation behind it. Additionally, patients may fear that this lack of transparency and eventually, dialogue, with new healthcare systems, may result in the misuse of their personal information and privacy, as they would not be aware of what is happening with their data and images.

The final topic of discussion in this chapter is the idea of healthcare practice and privacy during COVID-19. It is hard to navigate the fine line of what is acceptable to share within a pandemic and what is not in terms of patient data. For example, if a front-line worker is taking care of a patient, would it be ethical for them to check the patients records to see if they are positive for COVID-19 or would that be crossing the line? According to our knowledge of everyday practice in the healthcare sector, nobody is supposed to access patient information if it is not provided consent for from the patient themself. However, in the middle of a global pandemic, to what extent do these practices hold true? Additionally, is this a conversation that has to happen to the public? Therefore, it is important to acknowledge the differences that occur in terms of patient privacy and confidentiality in the midst of a pandemic as opposed to everyday practices. This will allow for a more open conversation about the idea of patient protection, which would let the general public have a better understanding of the appropriate standards.

To summarize this chapter, there are many perspectives to consider when looking at the privacy and security concerns surrounding the implementation of automation in the healthcare industry. The first one that was explored was the idea of online transmission and the potential risks that these poses. This includes the transfer of medical and financial information, as well as interactions with a healthcare professional. Hackers can try to gain access to this information and misuse this data, while fraudsters can try to get access to financial information. Thus, in order to prevent this, there are many precautions that can be put in place -- both on the end of the healthcare facilities and the patients themselves. This includes having a secure connection to reliable network, complex passwords, and two-factor encryption. Additionally, there are many ethical issues that arise with automated healthcare. This includes the idea of transparency and how automated healthcare can still be transparent to their patients and the notion of privacy and confidentiality guidelines during normal times when compared to a pandemic. Looking at what is acceptable and what is not is an open conversation to have, and this discussion would help contribute to the awareness of these issues to the general public. To conclude this topic, successful implementation of automated healthcare is definitely possible, however, it must occur with awareness and education about privacy and confidentiality.

Advancing Artificial intelligence in the fight against COVID 19

Introduction

The COVID-19 pandemic has certainly been one of the biggest international crises we have faced in the past century, affecting over 16.5 million people by the end of July 2020 (Worldometer, 2020). Due to this health crisis, research institutes, organizations, and companies around the world have invested time and resources in developing innovative solutions to solve the problems our healthcare systems are currently facing. In fact, in the first eight days of 2020, almost 500 artificial intelligence articles were published about coronavirus in hopes that such efforts would help develop a vaccine or other effective treatments to combat COVID-19 (Ulhaq, Khan, Gomes, & Paul, 2020). With these solutions and a plethora of data comes advancements in artificial intelligence in fields such as social network analysis, computer vision, and natural processing language. As previous chapters have already extensively covered the robotics and artificial intelligence that is involved directly in the healthcare industry such as diagnosis, infection detection, sample collection, and nursing, this chapter will focus on the advancements in artificial intelligence technology on indirect fights against COVID-19 such as predicting spread, enforcing social distancing, and developing tools to help scientists more effectively create machine learning models.

Advancements in Computer Vision Technology

Computer vision is a subfield of artificial intelligence that studies how computers can extract high-level understandings from digital images, with it making a significant breakthrough in recent years in the medical imaging industry (Ulhaq, Khan, Gomes, & Paul, 2020). However, during the pandemic, there have been advancements in computer vision to help enforce social distancing. The use of masks has been promoted as a predominant method to prevent the spread of COVID-19, with Centers for Disease Control and Prevention (CDC), one of the United States' national public health institutes, recommending it as a key prevention measure (Centers for Disease Control and Prevention, 2020).

With the release of the Masked Face Detection Dataset (MFDD), Real-world Masked Face Recognition Dataset (RMFRD) and Simulated Masked Face Recognition Dataset (SMFRD), many researchers have raced to implement computer vision artificial intelligence models to accurately detect whether individuals are wearing masks (Hariri, 2020). Experimenting with different numbers of artificial intelligence neurons ranging from 50 to 90 different neural networks, computer vision systems were able to detect masked faces with a 91.3% accuracy using Python libraries such as OpenCV and TensorFlow (Hariri, 2020). Other researchers from China have developed a model with 95% accuracy (Ulhaq, Khan, Gomes, & Paul, 2020), with expectations that these models will only continue improving. Companies such as Leeway Hertz have already launched these models into production, integrated with real-time monitoring dashboards and notification features (Leeway Hertz, 2020). These integrated features will help maintain individuals' privacies and will ensure that only authorities will have access to such data to maintain confidentiality.

In fact, countries have already started to implement these measures. In France, this AI powered tool developed by DakataLab has already been used in high-density tourist areas such as Cannes and France since late April 2020, which has enabled authorities to have statistics on the percentage of people that are wearing masks (Hadvas, 2020). However, in addition to enforcing social distancing rules in France where it is mandatory to wear masks on public transport systems, authorities are using these models to help understand the public's needs, as France is transitioning to distribute free masks to all residents (Hadvas, 2020). Understanding the percentage of people who wear masks will help determine how many masks the government will need to distribute to its residents.

Advancements in Social Network Analysis

Historically, artificial intelligence has been used before to track the spread of infectious diseases previously, with the 2009 H1N1 flu outbreak being one of the most recent examples. Using Google search terms and software from Google Flu Trends, researchers in the United States were able to predict the next regions affected by the outbreak using correlations between these search terms, giving local authorities the ability to prepare for the outbreak and mitigate the effects of that given region (Mayer-Schönberger & Cukier, 2017).

However, with the recent rise of social media users, researchers think there is now a better way to predict such outbreaks. With billions of users, incorporating social network analysis into their prediction models may prove to have increased accuracy, as social media messages and activity are a strong indicator of individual behaviour (Carey, 2020). Northeastern University's Network Science Institute's director Alessandro Vespignani is currently leading a project that uses cloud computing and census data to generate a dense collection of interconnections, mapping each individual's interactions with other people in the community (Carey, 2020). Researchers can then use social media messages to predict where an individual might travel next, gaining the ability to track an entire community's movements. Additionally, their model tracks activity such as a significant increase in search terms relating to COVID-19 such as "fever", "pneumonia", or "coronavirus". The research institute's next step is to incorporate these search term data into their dense interaction mapping system (Carey, 2020). This combination would make a powerful tool to predict the spread and the extent of COVID-19 in specific regions in the United States, and with enough time, the entire world.

Furthermore, social network analysis has been performed to

evaluate the effectiveness of public users' influence on social media. This research report analyzes the Twitter movement of 2,864 public figures such as Barack Obama and Donald Trump, helping efforts to determine which public users have the most significant sway on the general public. Most significantly, this research article finds that the most influential public figure is Donald Trump, with his influence equating to the sum of the influences of the next fourteen most influential public figures (Yum, 2020). Other significant public users include Barack Obama, the World Health Organization (WHO), and the regional offices of CDC (Yum, 2020). Research studies such as these show that the most significant public figure during COVID-19 is normally the political leader of a country such as President Donald Trump, despite the expertise of health organizations such as CDC and WHO. Such information is vital for authorities to understand how to handle current and future crises and infectious disease outbreaks, as it shows who the general public turns to during times of crises.

Advancements in Natural Language Processing

Natural language processing is a branch of artificial intelligence that studies the interaction between computers and natural language and analyzing the human language in a manner that is valuable. Such fields have had massive breakthroughs during COVID-19, mainly due to its widespread applications. For instance, Columbia University, led by research scientist Smaranda Muresan, has started a project investigating the impact of COVID-19 on individuals' mental health, citing an expected increased rate of depression due to job losses and economic downturn (Columbia Research, 2020). Through natural language processing, this tool will predict whether individuals are more susceptible to be affected by such changes. This shows that artificial intelligence will not only be of use to help fight COVID-19 during the pandemic, but tools can also be developed to prepare for the lasting economic and social impacts on our communities.

Another natural language processing project will help researcher's filter and find relevant research papers. The National Institute of Health (NIH)'s COVID-19 portfolio reported over 28,000 recent articles on the coronavirus, making it extremely difficult for researchers to currently read and find relevant papers (Huston, 2020). These natural language processing tools can help not only find relevant papers, but may also be able to extract relevant findings, saving researchers significant time in reading relevant literature. Through this tool, researchers will be able to spend this saved time on further refining and developing machine learning models, which would in turn have a greater impact on advancing artificial intelligence in the fight against COVID-19.

One application of natural processing language that will surely be adopted by more countries as COVID-19 progresses is using speech recognition to answer the public's most popular questions concerning COVID-19. In addition, their chatbot can help determine from a series of questions whether an individual has COVID-19. In France, AlloCovid tracks and transcribes conversations, as well as the respiratory and breathing pattern of a caller to determine the appropriate path of medical care for an individual (Business Solutions Atlantic France, 2020). This technology has been deployed to serve over one thousand people at one time and has already been deployed to other countries such as the United States, helping to unburden the hospital efforts by having individuals self-diagnose themselves through this chatbot (Business Solutions Atlantic France, 2020).

Advancements in Deep Learning

Deep learning is another field of artificial intelligence that studies a machine's ability to mimic the human brain in processing data. These advancements have particularly helped researchers understand and predict the future of this pandemic. Researchers have created a website that effectively predicts the number of outbreaks by country and region, with some predicting by November 2020, 10% of the American population would have been diagnosed with COVID-19 (Gu, 2020). In addition to predicting infection rates, such models can also predict the number of deaths by region and country with high accuracy.

Although these projections help policy makers and physicians understand the scope of COVID-19, deep learning has found many other applications during COVID-19. Deep learning has been able to identify high-risk patients before they even develop the virus (Wang et al., 2020). Using patient data from seven hospitals of approximately 7,000 patients, this deep learning system was able to differentiate the predictions of getting COVID-19 from other respiratory diseases such pneumonia with 88% to 89% accuracy (Wang et al., 2020). This powerful tool will help in areas where there is a shortage of testing kits and a lack of medical equipment. By mitigating the screening burden, the spread of COVID-19 will decelerate, as it will help to identify which at-risk persons should isolate immediately (Wang et al., 2020). Although the accuracy of the deep learning model could be increased, this tool suggests methods to prevent the infection from ever occurring based on a variety of risk factors an individual may be present with. Such risk factors may include age, respiratory conditions such as asthma, or other health complications such as obesity and high blood pressure.

Conclusion

COVID-19 has produced a plethora of data, from coronavirus case numbers to hospital records to tweets, and with this huge quantity of data, researchers, organizations, and companies have tried to implement innovative solutions to solve specific problems faced by the healthcare industry. Evidently with the rise in the number of research papers published, especially relating to coronavirus, there have been massive outbreaks relating to artificial intelligence. Many efforts have contributed to robotics and artificial intelligence systems in the healthcare industry pertaining to more effective diagnostics, nursing, sanitation, and sample collection. However, there have been several breakthroughs in other areas of artificial intelligence that may indirectly impact on COVID-19, such as in the respective fields of computer vision, natural language processing, speech recognition, and deep learning. With these breakthroughs, it has shown that in an effort to combat against COVID-19, we have turned to innovative solutions. Although a lot of these solutions are under development and are still being researched, a select few of these technologies have already been deployed and are currently used in countries such as France, China, and the United States.

As the pandemic continues, it is only expected that the artificial intelligence research and developments will only increase, and more innovative solutions will be developed to fight COVID-19. However, as discussed, the fight against COVID-19 is not only through direct artificial intelligence models such as computerized technology (CT) scans. The fight includes all encompassing fields such as the development of tools that help researchers implement machine learning models more quickly and studies to help individuals who will be more affected by the economic and social impacts of COVID-19. It will also provide insights on how to more effectively handle future infectious disease outbreaks or other related crises.

Post COVID 19 impacts on automation — harder to go back after robots take over

Having observed the landscape of automation prior to and during COVID-19, it becomes necessary to discuss how automation will affect humanity's transition into a post-COVID-19 era. American politician Henry Kissinger who served under the presidential administrations of Richard Nixon and General Ford, predicts that the pandemic will cause a global economic recession, far worse than the Great Depression. This will in turn, affect the nature of international relations that were established after the end of the Cold War (Shafiev, 2020). Other prominent politicians and scholars argue that the accelerated implementation of automation will likely foster economic prosperity, but come at the cost of increased inequality. The important question now becomes: how can we support humans in their day to day jobs with automation instead of how to replace them with it (Bloom & Prettner, 2020). LearnBonds, a financial news company, presented accumulated data that predicts the AI market value will jump from $22.6B to $126B by 2025. Furthermore, current AI market size by region shows that North America is the leading AI software industry in the world, followed by Asia-Pacific, and lastly, Europe (Furrer, 2020). These markets are going to continue to increase, prompting an effective response from each country to aid its citizens effectively cope with changes. These responses may include looking at how various industries and the blue collar workers within them will be affected. Additionally, healthcare systems have seen an unprecedented amount of innovation with regards to the growth of telehealth and robotics assisting front-line workers. An effective government response will include amendments to policies within healthcare, industry, and travel with relation to automation to ensure that a large majority of the population doesn't become unemployed and can still show up to their workplace. Lastly, automation will only add to the psychological changes that accompany social isolation for an extended period of time and contribute in fostering anxiety about close contact.

Automation has created an inherent problem in various industries, threatening the overall job market, prompting many to consider alternative career options once COVID-19 is done and over with. A 2019 Oxford Economics report predicted that 12.5M manufacturing jobs will be automated in China by 2030. Similarly, countries in Asia like South Korea have 7 robots per 100 workers, with every third robot being installed in China. The International Federation of Robotics (IFR) has reported that the cost of robots has been decreasing and will continue to decrease, allowing more companies and organizations to implement them into their practices (Farshchi, 2020). It is important to note that there still exist some jobs that require high competition for labour and talent like mechanics, technology repair technicians, delivery/postage-related drivers, etc. A post-COVID-19 era is still promising in terms of job security regarding these industries and although they may introduce some automation, the large majority of tasks will require the need of humans. A reason as to why many of these industries, especially retail and manufacturing, tend to draw less attention is because the large majority of individuals talented in technology gravitate towards areas like Silicon Valley, taking their technical expertise with them. This ties into the idea that high income jobs are not only harder to automate, but less affected by COVID-19, as opposed to low income jobs (<$20/hr) that can readily be automated. In fact, many warehouses across the globe have become completely autonomous, called dark warehouses, and require no human interaction whatsoever (Speer, 2020). Professor Joel Blit of the University of Waterloo's Department of Economics used health related incentives to rank industries (average physical proximity of workers) as well as the feasibility of doing so (fraction of workers in routine operations). He concluded that the retail sector is most likely to undergo an economic transformation, accompanied by wholesale, construction, and transportation (Media Relations, 2020).

Besides manufacturing and warehouses, other industries that

have been severely affected by the COVID-19 pandemic are movie theatres and airlines. Prominent theatre companies like Cineplex Entertainment have completely shut down all theatres across Canada to prevent the spread of coronavirus and promote social isolation. On June 3, 2020 another Cinema chain, AMC Theatres reported to CBC that they may not survive the pandemic. Specifically, the company mentioned "that it had enough cash to reopen its theatres this summer, as it had plans to do so, but if it's not allowed to open, it will need more money, which it may not be able to borrow." Even after theatres eventually open back up, many people may still watch from online, as they don't want to sit in crowded spaces in fear of the virus. To add on, admissions have been slowly declining since 2005 for theatres in Canada and the U.S., meaning that the industry will continue to suffer great losses for the foreseeable future, unless they are able to come up with a plausible solution (Arbel, 2020). This may come in the form of implementing AI to entice audiences. Movie theatres are only starting to catch up to Netflix and other competitors with regards to access to customer information. Chief Executive of Movio Will Palmer stated that "Once you understand that breakdown of how likely someone is to watch a film, you can then change the type of motivation that you put in front of them to come to the movie." Implementing AI with regards to the data analytics and campaign management software that Movio offers, could be a gamechanger that turns the tide for movie theatres (Erbland, 2019).

Airlines have suffered detrimental losses to the COVID-19 pandemic with virtually all countries issuing a non-essential travel ban to prevent the spread of the virus across borders. Similarly to movie theatres, many consumers may be less inclined to board a flight, as it is too crowded incite fear of being exposed to the virus. This could make the transition for pilots and other crew members extremely difficult and risk their job security, even after travel is permitted. With regards to automation, many commercial airlines that were in service prior to the coronavirus, have autopilot installed, but other than that, the general consumer isn't really aware of anything else that may be automated. However, the general perception on automation being implemented in aviation is not very good, as many have looked towards the role of automated systems in crashes. For example, investigations on both the 2018 Lion Air Flight 610 and 2019 Ethiopian Airlines Flight 302 crashes have probed the possible role of automated systems. When it comes to highly automated systems, the pilots can lose track of what is actually happening. This is allegedly what occurred in 2009 when Air France Flight 447 crashed in the Atlantic Ocean. "Airspeed sensors failed, causing the autopilot to turn itself off, but the pilots weren't able to figure out what was happening or how to recover" (LaBarre, 2019). However, the industry is in support of automation and many airlines suggest that it would save money and mitigate the problem regarding a shortage of qualified pilots. Even if airlines suggest that a digital transformation is necessary for the recovery of the air transport industry, in the end, it's up to the consumer if he or she wishes to fly in an autonomous aircraft.

Telehealth refers to receiving healthcare remotely and due to COVID-19, consumer adoption has catapulted from 11% of US consumers using telehealth in 2019 to 46% consumers using it now. With non-critical healthcare visits being cancelled, telehealth providers are seeing 50 to 175 times the number of patients than they did before, prompting them to quickly scale up offerings. Prior to COVID-19, US telehealth vendors were worth an estimated $3B, with larger ones prioritizing "virtual urgent care", which involves aiding consumers in getting instant telehealth visits with physicians (who they likely don't have a relationship with). With COVID-19 rapidly accelerating the expansion of Telehealth, "up to $250B of current US healthcare spend could potentially be virtualized." These services are set to be in place for at least the next 12 to 18 months, until a vaccine is available for mass distribution. As time passes, consumers will realize what their preferences are in terms of receiving healthcare, potentially leading virtual health to be integrated into the current care delivery system. This means that in the post-COVID-19 age, many consumers may choose to remain at home and participate in virtual healthcare visits which could change the whole regulatory landscape of healthcare across the world. However, there is currently a gap between consumer interest in telehealth (76%) and the actual usage (46%). Companies will need to increase awareness of telehealth offerings and educate the public on what exactly can be offered as well as the specifics around insurance coverage (Bestsennyy, Gilbert, Harris, & Rost, 2020). These changes will be felt on both sides, as doctors and medical student curriculums may start teaching how to interact with patients through virtual mediums.

The key to an effective transition that allows for employees to return to the workforce with relatively little inconveniences may come through policy reform. There will likely be quite a lot of jobs that have been automated during COVID-19 that will remain automated post-COVID-19, meaning that their human predecessors will have to look into other career opportunities. Professor Joel Blit states that "Policies like the Canada Emergency Wage Subsidy are stifling this economic transformation and need to be phased out. Instead, we should focus on supporting workers through the transition." He suggests that current programs like the Canada Emergency Response Benefit (CERB) program be transformed into a universal basic income to help alleviate the burden of temporary unemployment. Furthermore, a program that helps to retrain employees could be coupled with CERB to refresh the memory of employees, so they can hit the ground working from day one (Media Relations, 2020). Similarly, the healthcare industry may need to develop new policies and guidelines regarding telehealth and virtual health check-ups to address issues like patient-doctor confidentiality (no screen recording) as well as general privacy and monitoring concerns. Having these guidelines in check will better prepare countries for a future pandemic and help mitigate the enormous strain on healthcare workers and facilities. Other policies may address automation directly and look into the ethicalities of human-automation hybrid jobs. For those who wish to work out of the country or attend international internships, separate policy amendments may need to occur with regards to immigration and making sure international relations are not disrupted.

Around early April, Canada sent "millions" of masks and other medical supplies to fight COVID-19, which caused an uproar from the public. Although the Chinese government has since repaid the amount for these supplies, it demonstrates the "battening down the hatches mentality." A large majority of people agree that borders should be strictly kept closed, as they argue that spread of pandemics primarily depends on the movement of people. U.S. President Donald Trump has abided by these beliefs and temporarily banned immigration to the U.S, causing travel to become heavily restricted. With escalating tensions between nations over medical supplies and immigration, many have voiced concern that a united global response, in which nations work together, is necessary (Omidvar , 2020). Liliia Khasanova, researcher at Stockholm University, states that "We can't go back to normal, because the normal that we had was precisely the problem." She suggests that there are two scenarios: cooperation and protectionism. To succeed in achieving minimum damages to the global economy, countries across the globe need to cooperate to form a "coordinated multilateral response" (Khasanova, 2020). In Canada, billions are being spent by all governments, municipalities, and provinces to aid people affected by job loss. An estimated $100 billion could be added to the national debt depending on how long the crisis lasts. On top of that, Canada's aging population warrants a tremendous amount of newcomers to fill jobs and stimulate the economy. If immigration is restricted, then it can be argued that the economy will suffer even greater losses (Omidvar , 2020).

An issue that ties into many industries including air transport, movie theatres, and retail is the reemergence of physical contact. With a lot of processes now being automated, some areas that the public go to visit, like retail clothing stores, may be void of any employees which could add onto the psychological effects of chronic social distancing. As mentioned before, much of the public may not immediately go back into crowded movie theatres or book an international flight due to social anxiety and fear of being exposed to the virus. Furthermore, this may bleed into movies and marketing, with scenes or advertisements containing a lot of physical contact to be omitted in an attempt to maintain viewership. Robin Dunbar, professor of evolutionary psychology at Oxford University says "Physical contact is part of the mechanism we use to set up our relationships, friendships and family memberships." Looking back at our primate ancestors, stroking hairy skin triggers the release of endorphins (chemicals that reduce stress), making us feel warm and happy. Overall communication might be slightly more complicated and some greetings or gestures may cease to be acceptable. During March, France's health minister recommended that citizens don't kiss because of COVID-19. It is interesting to note that during the 15th century in England, King Henry VI similarly banned kissing to limit the spread of the bubonic plague (Walker, 2020).

Automation will continue to have tremendous impacts on industry even after the COVID-19 pandemic ceases. Based on analysis from professors and other experts, the retail industry is likely to undergo the greatest economic transformation, as many have become accustomed to online shopping and many warehouses have become 100% autonomous. Retail is closely followed by wholesale, construction, and transportation industries. Additionally, movie theatres chains are attempting to transition towards automated processes to entice audiences, as COVID-19 has proved to have disastrous economic implications on these facilities. The air transport industry is also trying to automate more processes, but will potentially face consumer draw-back due to limited public knowledge of the extent of automation within these aircrafts. With regards to healthcare, virtual health care visits and the telehealth industry have skyrocketed in terms of membership and will likely prompt the integration of some virtual services in current healthcare systems. The effect of automation on all of these industries correlates with the psychological impact of social isolation on social interaction and physical communication. A large majority of the population may initially present symptoms of social anxiety when faced with physical interactions and greetings. A highly emphasized point throughout this chapter is that governments across the globe should work together and develop a unified response to help their citizens transition back to some semblance of normalcy instead of closing their borders and independently trying to deal with the pandemic.

Conclusion

The term "artificial intelligence" (AI) was first coined in 1956 by John McCarthy (Amisha, Malik, Pathania, & Rathaur, 2019). However, despite its brief popularity in the research field in the 1980s and early 1990s, companies and research institutes did not seriously invest time into AI research until 1997, after IBM's Deep Blue became the first computer to defeat a chess grandmaster (Amisha, Malik, Pathania, & Rathaur, 2019). Ever since then, its applications have since been extensively researched in many industries, including its use in our healthcare system. In fact, in 2016, the largest portions of investments in AI research went to its applications in healthcare automation (Amisha, Malik, Pathania, & Rathaur, 2019), so it is without a doubt that within the next decade, major medical breakthroughs will be made. Looking back at AI's developmental history is crucial to understand the impact it has had on our fight against COVID-19. Although AI cannot be used to solve all of our healthcare industry's current issues, it is clear that researchers have made profound progress within just a little over twenty years. AI has been able to handle many healthcare tasks such as sample collection and medical diagnosis, and it is becoming increasingly clear how powerful AI really is because it has automated tasks in the healthcare industry that were not previously possible. With its developments in the COVID-19 pandemic, AI has established its place in our healthcare system not only in our efforts to fight this pandemic, but also across hospitals in the next few decades.

The COVID-19 pandemic has devastated millions worldwide and will have severe economic consequences for the upcoming years. During these times, there has been a high demand for medical care and products. However, with the staggering number of cases of COVID-19, the former has been difficult to meet since there are simply not enough doctors or hospital equipment to oversee all those that are infected. In Italy, there were reports that they had run out of medical equipment and were starting to have to choose between those who would receive treatment and those who would not, prioritizing those who had the greatest chances of survival, which lead to a higher death toll rate (Beall, 2020). Other issues presented during the pandemic was insufficient production of testing kits, broken communication systems between patients and companies for their medication, and the psychological impacts the pandemic has had on frontline workers and patients. The pandemic presented clear needs that needed to be addressed and within a few weeks, scientists have transformed AI into technological tools that can aid frontline workers in our fight against COVID-19. The development of such technologies and research has never been so quick and so innovative, which has allowed us to decrease the number of COVID-19 cases and has helped patients meet their medical needs despite limited hospital spaces and the lockdown restrictions. Such technologies, as presented in Chapter 2, include OmniSYS's interactive voice system to help coordinate the pharmaceutical companies' delivery of patient medicine and essential services (OmniSYS, 2020).

The need for healthcare automation using robotics has been made abundantly clear in terms of helping decrease the burden of the responsibilities of frontline workers. First, it has streamlined the process of diagnosing COVID-19 and its responses in sanitation, sample collection, and infectious disease prevention, with all technologies focusing on protecting workers. Since the beginning of COVID-19, significant headway has been underway in diagnosing patients, with machines being able to diagnose COVID-19 from pneumonia with 95% accuracy within just three seconds (Bahl et al., 2020). Additionally, chapter 3 has also explored companies' races to produce more testing kits and faster times in diagnosis with these testing kits. Technologies have certainly made this possible with companies testing up to one thousand to three thousand tests a day and individuals knowing the results of their tests within twelve to twenty four hours (Sanders, 2020).

One major problem that frontline workers were facing was sample collection because it increased their chances of becoming infected due to the closeness to those who were possibly infected and the volume of patients they were testing in one day. As presented in chapter 5, many robots have since been deployed in hospitals for viral testing, including a throat swabbing robot from Denmark that can perform this operation in twenty to thirty seconds (Crowe, 2020). These technologies are especially essential to protect our frontline workers, as it decreases the contact time they would have with potentially infected samples. Moreover, the development of such technologies can help the sample collection process to become more efficient, as robots are able to automatically perform these actions. Certainly, this type of technology has been a major breakthrough in protecting frontline workers and will be expanded to be used in any future infectious diseases.

AI and similar technologies have been vital in other fields, such as its use in infection prevention and sanitation. While UV-C technology has already been extensively applied, companies worldwide have been developing similar robotics specifically customized to fight COVID-19. For instance, robots such as Xenex Lightstrike have been specifically created to clean hospital tools, as previously used robotics were not able to effectively clean small spaces (Sascha, 2020). This has reduced the responsibilities of frontline workers, as they are no longer required to sanitize hospital rooms and equipment. The development of this technology has also helped decrease the time that frontline workers are in close contact with infected areas, and will help tremendously in protecting our frontline workers. However, such technology is not only helpful in the healthcare system, but will also reduce infections in the general public, with Miso Robotics developing a thermal screening system at restaurants. By detecting people with high fevers or other COVID-19 related symptoms quickly, they will help decrease the number of COVID-19 cases in the general public, as discussed in Chapter 7.

Other diagnosis techniques and systems are introduced in Chapter 4, discussing the widespread applications of intelligent theromemtry systems in the past few decades. During the COVID-19 crisis, these intelligent systems have been adapted to fight COVID-19. For instance, Kinsa has implemented a smart thermometer application that detects COVID-19 hotspots, aiding authorities on identifying the next high-risk areas and helping with strategic responses to help those areas (Indelicato, 2020). While agencies have expressed concerns in using these systems to diagnose COVID-19 due to its likelihood of inaccurately diagnosing patients, it has shown progress in becoming a system that can predict the outbreak. If this model proves to be successful, these thermometry systems will help predict future infectious disease outbreaks before they even happen and prevent further pandemics.

Other technologies have helped decrease the burden placed on nurses during COVID-19. Due to the high volumes of patients, nurses have had an incredible burden taking care of so many patients. However, technologies such as Tommy the Robot nurse robot in Italy as discussed in Chapter 8, can help decrease the responsibilities for the nurse so that they can more effectively take care of larger volumes of patients during this pandemic. Its microphone and tablet allows for efficient communication between the physician and the patient, and its ability to monitor essential health metrics such as oxygen saturation has made it a vital technology (). The implementation of this robot has already seen great success in reducing the number of cases and could be used worldwide, especially in regions with higher numbers of cases and thus higher burdens on the country's healthcare system.

However, there have also been extensive technologies that have not involved directly interacting with the patient or frontline workers. These technologies have been used to advance AI, but have played a more indirect role during the COVID-19 pandemic. Several breakthroughs in computer vision, natural language processing, and social network analysis have provided technological tools that can further advance our ability to fight COVID-19. As discussed in chapter 14, computer vision models have been deployed to help authorities enforce social distancing regulations and understand how to distribute their resources to those most in need (Hadvas, 2020). Social network analysis has been able to track the possible spread of COVID-19 and use social media to more effectively predict the outbreak (Yum, 2020). All of these technologies have had a crucial, yet indirect impact on our ability to control COVID-19, but also to prevent another pandemic on this scale from ever happening again.

Despite these breakthroughs in technologies, it is vital to analyze the cost, efficiency, economic, and psychological impact of these technologies to effectively transform these models from development to production. Without a doubt, these technologies have made tasks more cost-friendly and efficient. For instance, normal testing that would take two to three days to process the results of the testing kits took robotic technologies less than twenty four hours (Sanders, 2020). As discussed in chapter ten, although the application of many of these technologies are quite expensive at first glance, its efficiency at getting tasks done have proven that these technologies are worthwhile to invest into.

However, there are other economic considerations to discuss beyond just its costs. Technology is much more efficient at performing tasks because they can be done without breaks and almost always can complete the task more effectively. Despite its efficiency and effectiveness, since many of these tasks are becoming automated, many people could potentially lose their jobs. This displacement of workers could have significant impacts on unemployment rates and people's earnings. Chapter 11 discusses the unemployment impacts of transitioning into healthcare automation, but it should be understood that the removal of these positions opens new ones to supervise these machines. The difficult part with this more than probable transition is transitioning these workers from performing these tasks manually to being able to use machinery and technology to automatically perform such tasks.

COVID-19 has also delved deeper into using such technology to help diagnose those with mental health disorders. Although this has become previously relevant, it will become increasingly crucial to investigate because there will be major psychological impacts from this pandemic on the general population. Chapter twelve further investigates the links between such technologies and analyzing the psychological impact. It is becoming vital to analyze these links because unlike many physical disorders or diseases, these mental issues cannot be physically or quantitatively measured. Similar psychological research has been focused during the COVID-19 such as Columbia University's research scientist's project investigating the psychological impacts due to the economic damages during COVID-19 (Columbia Research, 2020).

COVID-19 has been one of the largest pandemics, in terms of scale, that we have faced in the past century. To fight it, we have developed and transformed many technologies to help automate healthcare tasks in order to protect frontline workers and to increase the chances of helping a patient recover from COVID-19 faster. A crisis to this scale has led to incredible innovation from scientists, research institutes, companies, pharmaceuticals, and other organizations from all over the world. From sanitization robots from Lithuania to nursing robots in Italy, the world has clearly embraced the use of AI to help mitigate the responsibilities of frontline workers, so they can focus on caring for the most critical of patients. The role AI has had on this pandemic speaks clearly to its future in the healthcare system, namely that it will have a crucial role in the future with a breakthrough most probably happening in the next decade or so. Despite the potential economic setbacks or psychological impacts, on the whole, AI presents a far greater advantage to society because while there may be short-term detriments, there is a far greater long-term benefit to investing in AI due to its undeniable efficiency and effectiveness. Although COVID-19 is expected to drag out for at least another year with the fear of subsequent waves, it is safe to say that AI will play a crucial role in mitigating these consequences and protecting our frontline workers. While the current pandemic presents scary challenges, AI has addressed effective solutions and by doing so, it has confirmed its place into our future healthcare systems.

Works Cited List

Acieta. (2020). Medical Robotics. https://www.acieta.com/why-robotic-automation/robotic-solutions-industry/medical/

Ackerman, E. (2020, March 11). Full Page Reload. IEEE Spectrum: Technology, Engineering,

and Science News. https://spectrum.ieee.org/automaton/robotics/medical-robots/autonomous-robots-are-helping-kill-coronavirus-in-hospitals

Admin, A. (2020, January 19). The economic effects of automation aren't what you think they are. Retrieved from https://www.singlelunch.com/2019/10/21/the-economic-effects-of-automation-arent-what-you-think-they-are/

AI in Healthcare: Data Privacy and Ethics Concerns. (2019, November 18). Retrieved from https://www.lexalytics.com/lexablog/ai-healthcare-data-privacy-ethics-issues

American Cancer Society. (2020, January 1). Ultraviolet (UV) Radiation. Ultraviolet (UV) Radiation. https://www.cancer.org/cancer/cancer-causes/radiation-exposure/uv-radiation.html

Amisha, Malik, P., Pathania, M., & Rathaur, V. (2019). Overview of artificial intelligence in medicine. Journal of Family Medicine and Primary Care, 8(7), 2328. doi:10.4103/jfmpc.jfmpc_440_19

Apec Water. (2017). How does ultraviolet water purification work? | APEC Water. Free Drinking Water. https://www.freedrinkingwater.com/whole-house/water-filter-knowledge-base/how-does-ultraviolet-water-purification-work.htm

Arbel, T. (2020, June 3). Cinema chain AMC warns it may not

survive the pandemic | CBC News. Retrieved August 4, 2020, from https://www.cbc.ca/news/entertainment/amc-warning-pandemic-finances-1.5597040

Automation and a Changing Economy: The Case for Action. (2019, May 17). Retrieved from https://www.aspeninstitute.org/publications/automation-and-a-changing-economy-the-case-for-action/

Bahl, S., Haleem, A., Javid, M., Suman, R., Vaish, A., Vaishya, R. (2020). Industry 4.0 technologies and their applications in fighting COVID-19 pandemic. Diabetes & Metabolic Syndrome: Clinical Research & Reviews. Volume 14(4), P. 419-422. Retrieved from https://doi.org/10.1016/j.dsx.2020.04.032

Balter, M. L., Leipheimer, J. M., Chen, A. I., Shrirao, A., Maguire, T. J., & Yarmush, M. L. (2018, June 1). Automated end-to-end blood testing at the point-of-care: Integration of robotic phlebotomy with downstream sample processing. PubMed Central (PMC). https://www.ncbi.nlm.nih.gov/pmc/articles/PMC6058193/

Banga, K. (2020, July 21). 3 ways technology can help global drug makers fight COVID-19. World Economic Forum. https://www.weforum.org/agenda/2020/07/3-ways-digital-technology-can-help-drug-makers-fight-covid-19/

Bauer, J., Hoq, M. N., Mulcahy, J., Tofail, S. A., Gulshan, F., Silien, C., . . . Akbar, M. M. (2020). Implementation of artificial intelligence and non-contact infrared thermography for prediction and personalized automatic identification of different stages of cellulite. EPMA Journal, 17-29. doi:10.1007/s13167-020-00199-x

BC Centre for Disease Control. (2020). Antibody Testing (Serology). Provincial Health Services Authority. http://www.bccdc.ca/health-professionals/clinical-resources/covid-19-care/covid-19-testing/antibody-testing-(serology)

Beall, A. (2020, April 20). The heart-wrenching choice of who lives and dies. Retrieved August 01, 2020, from https://www.bbc.com/future/article/20200428-coronavirus-how-doctors-choose-who-lives-and-dies

Bestsennyy, O., Gilbert, G., Harris, A., & Rost, J. (2020, June 1). Telehealth: A quarter-trillion-dollar post-COVID-19 reality? Retrieved August 4, 2020, from https://www.mckinsey.com/industries/healthcare-systems-and-services/our-insights/telehealth-a-quarter-trillion-dollar-post-covid-19-reality

Business Solutions Atlantic France. (2020). Allo-Media uses speech recognition and AI to detect COVID-19 symptoms. Retrieved from https://www.business-solutions-atlantic-france.com/news/allo-media-uses-speech-recognition-and-ai-to-detect-covid-19-symptoms/

Bloom, D., & Prettner, K. (2020, June 25). COVID-19 and the macroeconomic effects of automation. Retrieved August 4, 2020, from https://voxeu.org/article/covid-19-and-macroeconomic-effects-automation

Bloudoff-Indelicato, M. (2020, April 02). This company claims its smart thermometer could help detect coronavirus hot spots faster than the CDC. Retrieved July 24, 2020, from https://www.cnbc.com/2020/04/02/this-smart-thermometer-could-help-detect-covid-19-hot-spots.html

CADTH. (2020, May 6). Infrared Temperature Devices for Infectious Disease Screening During Outbreaks: Overview of an ECRI Evidence Assessment [PDF]. Ottawa: https://cadth.ca/sites/default/files/covid-19/ha0004-non-contact-ir-temperature-screening-final.pdf.

Carey, Benedict. (2020). Mapping the social network of coronavirus. Retrieved July 27, 2020 from https://www.nytimes.com/2020/03/13/science/coronavirus-social-networks-data.html

CBC News. (2020 Mar 14). South Korea praised for handling of COVID-19 Outbreak [video]. Youtube. Retrieved from https://www.youtube.com/watch?v=g-r79RXboXc

Center for Devices and Radiological Health. (2020, June 19). Non-contact Temperature Assessment Devices During COVID-19. Retrieved July 17, 2020, from https://www.fda.gov/medical-devices/emergency-situations-medical-devices/non-contact-temperature-assessment-devices-during-covid-19-pandemic

Centers for Disease Control and Prevention. (2020). Considerations for Wearing Cloth Face Coverings. Retrieved July 27, 2020 from https://www.cdc.gov/coronavirus/2019-ncov/prevent-getting-sick/cloth-face-cover-guidance.html

Centers for Disease and Control (2020, May 13). Symptoms of Coronavirus. Retrieved July 17, 2020, from https://www.cdc.gov/coronavirus/2019-ncov/symptoms-testing/symptoms.html

Chakraborty, A. (2020, May 08). Researchers test wastewater to identify Covid-19 hotspots. Retrieved from https://www.water-technology.net/news/researchers-wastewater-covid-19-hotspots/

Chappelow, J. (2020, January 29). Economic Growth Definition. Retrieved from https://www.investopedia.com/terms/e/economicgrowth.asp

Chaturvedi, A., & Anurag Chaturvedi co-leads Dasra's work on urban sanitation. (n.d.). Fixing India's Sewage Problem (SSIR). Retrieved from https://ssir.org/articles/entry/fixing_indias_sewage_problem#:~:text=Untreated sewage is the leading,which mosquitoes and germs breed.

Cipla. (2019). Manufacturing | Cipla. https://www.cipla.com/about-us/manufacturing

Clifford, C. (2020, March 23). Look inside the hospital in China where coronavirus patients were treated by robots. CNBC. https://www.cnbc.com/2020/03/23/video-hospital-in-china-where-covid-19-patients-treated-by-robots.html

Collins S.H., Dario, D., Choset, H., et al. (2020). Combatting COVID-19 The role of robotics in managing public health and infectious diseases. Vol 5(40). Retrieved from https://doi.org/10.1126/scirobotics.abb5589

Columbia Research. (2020). Natural language processing for understanding mental health around Covid-19. Retrieved July 27, 2020 from https://research.columbia.edu/covid/community/languageprocessing

Covid Tracking Project. (2020). US Historical Data. Retrieved from https://covidtracking.com/data/us-daily

Crowe, S. (2020, My 29). Throat swabbing robot developed for COVID-19 testing by Danish startup. The Robot Report. https://www.therobotreport.com/danish-startup-develops-throat-swabbing-robot-for-covid-19-testing/

Dalgaard, B. (2020, May 27). Robot swabs patients' throats for Covid-19. SDU. https://www.sdu.dk/en/nyheder/forskningsnyheder/robot-kan-pode-patienter-for-covid-19

Davenport, T., & Kalakota, R. (2019, June). The potential for artificial intelligence in healthcare. Retrieved from https://www.ncbi.nlm.nih.gov/pmc/articles/PMC6616181/

Davis, J. (2019, August 02). The 10 Biggest Healthcare Data Breaches of 2019, So Far. Retrieved from https://healthitsecurity.com/news/the-10-biggest-healthcare-data-breaches-of-2019-so-fr

Demaitre, E., Demaitre, E., Says, J. B., Bliss, J., Says, G. M., Modric, G., ... Kennedy, T. (2020, June 11). UVD Robots responds to surging demand during COVID-19 crisis. Retrieved from https://www.therobotreport.com/uvd-robots-responds-surging-demand-during-covid-19-crisis/

Demaitre, Eugene. (2020). Keenon rolls out disinfection robot. Retrieved from https://www.therobotreport.com/keenon-rolls-out-disinfection-robot-china-covid-19/

Dong, A. (2017, November 01). Elderly Care in the Age of Machine and Automation. Retrieved July 31, 2020, from https://blog.petrieflom.law.harvard.edu/2017/11/01/elderly-care-in-the-age-of-machine-and-automation/

Economic Times. (2020). Mangaluru company develops UV disinfection robot. Retrieved from https://economictimes.indiatimes.com/news/politics-and-nation/covid-19-mangaluru-company-develops-uv-light-disinfection-robot-which-sanitises-entire-room-in-4-minutes/videoshow/75717325.cms

Eising, P. (2017, December 07). What exactly IS an API? Retrieved

July 24, 2020, from https://medium.com/@perrysetgo/what-exactly-is-an-api-69f36968a41f

Erbland, K. (2019, April 2). Theaters May Use AI to Entice Audiences, From Ticket Pricing to Concessions. Retrieved August 4, 2020, from https://www.indiewire.com/2019/04/artificial-intelligence-movie-theaters-tickets-1202055135/

Farshchi, S. (2020, April 10). Expect More Jobs And More Automation In The Post-COVID-19 Economy. Retrieved August 4, 2020, from https://www.forbes.com/sites/shahinfarshchi/2020/04/10/expect-more-jobs-and-more-automation-in-the-post-covid-19-economy/

Flow Robotics. (2020). "Robot automates COVID-19 testing". Healthcare In Europe. Retrieved from https://healthcare-in-europe.com/en/news/robot-automates-covid-19-testing.html

Fluid Robotics: Detecting COVID-19 Outbreaks Through Wastewater. (2020, June 29). Retrieved from https://unltdindia.org/2020/06/22/fluid-robotics-detecting-covid-19-outbreaks-through-wastewater/

Fraud Alert: COVID-19 SCAMS: Office of Inspector General: U.S. Department of Health and Human Services. (2020, July 07). Retrieved from https://oig.hhs.gov/coronavirus/fraud-alert-covid19.asp

Fretty, P. (2020, March 31). StackPath. Industry Week. https://www.industryweek.com/technology-and-iiot/article/21127428/will-covid19-create-new-robotic-norm

Furrer, T. (2020, June). Artificial Intelligence & Automation in The Post COVID-19 Era. Retrieved August 4, 2020, from https://www.plugandplaytechcenter.com/resources/artificial-intelligence-automation-post-covid-19-era/

Genovese, A. M., & Minutes, 2. (2019, October 01). Top 5 cyberattacks against the health care industry. Retrieved from https://www.stormshield.com/news/top-5-cyberattacks-against-the-health-care-industry

GM to Make Face Masks for Frontline Workers. (2020, March

31). Retrieved from https://media.gm.com/media/us/en/gm/ home.detail.html/content/Pages/news/us/en/2020/mar/0331-coronavirus-update-7-masks.html

Gomes, D., Khan, A., Paul, M. & Ulhaq, A. (2020, May 05). Computer Vision For COVID-19 Control: A Survey. Retrieved July 27, 2020, from https://arxiv.org/abs/2004.09420

Grounds, M. (2020, March 30). Covid-19 Crisis drives demand for Automation in Healthcare – But at What Cost? Retrieved July 24, 2020, from https://www.globalpolicyjournal.com/ blog/30/03/2020/covid-19-crisis-drives-demand-automation-healthcare-what-cost

Gu, Youyang. (2020). COVID-19 Projections Using Machine Learning. Retrieved July 27, 2020 from https://covid19-projections.com

Hadavas, Chloe. (2020). France is using A.I. to detect whether people are wearing masks. Retrieved from https://slate.com/ technology/2020/05/france-artificial-intelligence-mask-detection-coronavirus.html

Hariri, W. (2020). Efficient Masked Face Recognition Method during the COVID-19 Pandemic. doi:10.21203/rs.3.rs-39289/v1

HealthLink BC. (2018, September 23). Eye Injuries Caused by Ultraviolet (UV) Light. https://www.healthlinkbc.ca/health-topics/ aa128596

HealthLink BC. (2020, July 24). Symptoms of COVID-19. https:// www.healthlinkbc.ca/symptoms-covid-19

Henderson, H. (2020). "IGI Launches Major Automated COVID-19 Diagnostic Testing Initiative". IGI News. Retrieved from https:// innovativegenomics.org/news/covid-19-testing-lab/

Hewitt, J. (2017, November 30). This is the hidden risk of automation that no one is talking about. Retrieved July 31, 2020, from https:// www.weforum.org/agenda/2017/11/automation-automated-job-risk-robot-bored-boredom-effort-fourth-industrial-revolution/

How Can OmniSYS Help Your Pharmacy Support COVID-19?

(2020, April 06). Retrieved from https://www.omnisys.com/how-can-omnisys-help-your-pharmacy-support-covid-19/

Huston, Matthew. (2020). Artificial intelligence tools aim to tame the coronavirus literature. Retrieved July 27, 2020 from https://www.nature.com/articles/d41586-020-01733-7

Indelicato, M. B. (2020, April 2). This company claims its smart thermometer could help detect coronavirus hot spots faster than the CDC. CNBC. Retrieved July 17, 2020, from https://www.cnbc.com/2020/04/02/this-smart-thermometer-could-help-detect-covid-19-hot-spots.html

Jacksonville University. (2020). The Difference Between Nurse Practitioners and Doctors. https://www.jacksonvilleu.com/blog/nursing/the-difference-between-nurse-practitioners-and-doctors/

Johnson, Khari. (2020). How people are using AI to detect and fight the coronavirus. VB. Retrieved from https://venturebeat.com/2020/03/03/how-people-are-using-ai-to-detect-and-fight-the-coronavirus/

Khan, Z.H., Lee, C.W., Siddique, A. (2020). Robotics Utilization for Healthcare Digitization in Global COVID-19 Management. 17(11), 3819. Retrieved from https://doi.org/10.3390/ijerph17113819

Khasanova, L. (2020, April 21). International Relations in the Post-COVID-19 Era: Cooperation Vs Protectionism. Retrieved August 04, 2020, from https://www.internationalaffairshouse.org/cooperation-vs-protectionism-in-post-covid-19/

Kopp, C. M. (2020, May 04). Globalization. Retrieved from https://www.investopedia.com/terms/g/globalization.asp

LaBarre, S. (2019, March 26). The future of aviation? Even more automation. Retrieved August 4, 2020, from https://www.fastcompany.com/90324699/the-future-of-aviation-even-more-automation

Lanco Integrated. (2020). "Making Diagnostic Test Kit Assembly Systems for COVID-19 On a Deadline". Robotic Industries Association. Retrieved from https://www.robotics.org/content-detail.cfm/Industrial-Robotics-News/Making-Diagnostic-Test-Kit-Assembly-Systems-for-COVID-19-On-a-Deadline/content_id/8940

Lee, B. (2020, February 16). For COVID-19 Coronavirus, How Well Do Thermometer Guns Even Work? Retrieved July 17, 2020, from https://www.forbes.com/sites/brucelee/2020/02/16/for-covad-19-coronavirus-how-well-do-thermometer-guns-even-work/

Leeway Hertz. (2020). Face mask detection system using artificial intelligence [Photograph]. https://www.leewayhertz.com/face-mask-detection-system/

Lessler, J., Azman, A. S., Mckay, H. S., & Moore, S. M. (2017). What is a Hotspot Anyway? The American Journal of Tropical Medicine and Hygiene, 96(6), 1270-1273. doi:10.4269/ajtmh.16-0427

Lupton, A. (2020, March 31). Thermometer shortage a source of frustration as COVID-19 cases rise. CBC News. Retrieved July 17, 2020, from https://www.cbc.ca/news/canada/london/covid-19-thermometer-shortage-1.5514855

Luthra, S. (2020, March 30). Temperature Check: Tracking Fever, a Key Symptom of Coronavirus. The New York Times. Retrieved July 17, 2020, from https://www.nytimes.com/article/coronavirus-temperature-fever-thermometer.html

Makhzani, A. (2020, May 21). Psychology of automation: How robots affect our mental wellbeing. Retrieved July 31, 2020, from https://www.worktechacademy.com/psychology-of-automation-how-robots-affect-our-mental-wellbeing/

Manufacturing Automation. (2020, July 17). Ontario PPE manufacturer launches automated surgical mask production. Retrieved July 24, 2020, from https://www.automationmag.com/ontario-ppe-manufacturer-launches-automated-surgical-mask-production/

Manufacturing Automation. (2020, May 19). Ontario manufacturer develops COVID-19 testing station to reduce use of PPE. Retrieved July 24, 2020, from https://www.automationmag.com/ontario-manufacturer-develops-covid-19-testing-station-to-reduce-use-of-ppe/

Manyika, J., Lund, S., Chui, M., Bughin, J., Woetzel, J., Batra, P., . . . Sanghvi, S. (2019, May 11). Jobs lost, jobs gained: What the future of work will mean for jobs, skills, and wages. Retrieved July 31, 2020, from https://www.mckinsey.com/featured-insights/future-of-work/jobs-lost-jobs-gained-what-the-future-of-work-will-mean-for-jobs-skills-and-wages

Marr, B. (2020, March 18). Robots And Drones Are Now Used To Fight COVID-19. Retrieved July 24, 2020, from https://www.forbes.com/sites/bernardmarr/2020/03/18/how-robots-and-drones-are-helping-to-fight-coronavirus/

Mayer-Schönberger, V., & Cukier, K. (2017). Big data: A revolution that will transform how we live, work and think. London: John Murray.

Mayo Clinic. (2020, May 20). How do COVID-19 antibody tests differ from diagnostic tests? https://www.mayoclinic.org/diseases-conditions/coronavirus/expert-answers/covid-antibody-tests/faq-20484429

Mayo Clinic Staff. (2018, September 15). Thermometers: Understand the options. Retrieved July 17, 2020, from https://www.mayoclinic.org/diseases-conditions/fever/in-depth/thermometers/art-20046737

Mayo Clinic Staff. (2019, September 11). Fever: First aid. Retrieved July 17, 2020, from https://www.mayoclinic.org/first-aid/first-aid-fever/basics/art-20056685

McGee, M. K., & Ross, R. (n.d.). Would More Telehealth Bring New Privacy, Security Concerns? Retrieved from https://www.careersinfosecurity.com/would-more-telehealth-bring-new-privacy-security-concerns-a-11208

Media Relations. (2020, July 31). Automation will transform work in the post COVID-19 economy - but are we ready? Retrieved August 4, 2020, from https://uwaterloo.ca/arts/news/automation-will-transform-work-post-covid-19-economy-are-we

Mitchell, D. (2020, April 21). Hamilton researchers hope 'robot colleagues' will help step up coronavirus testing. Global News. https://globalnews.ca/news/6845693/hamilton-robots-coronavirus-testing-covid-19/

Murphy, R. R. (2020, April 22). Robots are playing many roles in the coronavirus crisis – and offering lessons for future disasters. The Conversation. https://theconversation.com/robots-are-playing-many-roles-in-the-coronavirus-crisis-and-offering-lessons-for-future-disasters-135527

Neustaeter, B. (2020). Canada's first disinfection robot being tested. Retrieved from https://www.ctvnews.ca/health/coronavirus/canada-s-first-disinfection-robot-being-tested-1.4918668

Omidvar, R. (2020, May 4). Will Canada be as open to immigrants after COVID-19? Retrieved August 4, 2020, from https://policyoptions.irpp.org/magazines/may-2020/will-canada-be-as-open-to-immigrants-after-covid-19/Palma, S. (2020, March 22). How Singapore waged war on coronavirus . Retrieved July 24, 2020, from https://www.ft.com/content/ca4e0db0-6aaa-11ea-800d-da70cff6e4d3

Partovi, S. (2019, December 4). AWS and Novartis: Re-inventing pharma manufacturing. Amazon Web Services. https://aws.amazon.com/blogs/industries/aws-and-novartis-re-inventing-pharma-manufacturing/

Patel, P. C., Devaraj, S., Hicks, M. J., & Wornell, E. J. (2018). County-level job automation risk and health: Evidence from the United States. Social Science & Medicine, 202, 54-60. doi:10.1016/j.socscimed.2018.02.025

Pettinger, T. (2019, November 12). Automation - benefits and costs. Retrieved July 24, 2020, from https://www.economicshelp.org/blog/25163/economics/automation/

Phone Soap. (2020). How does UV light sanitize? Retrieved from https://www.phonesoap.com/pages/faq-uv-c-light-technology

Price, S. (2019, November 19). AI in psychiatry: Detecting mental illness with artificial intelligence. Retrieved July 31, 2020, from https://www.healtheuropa.eu/ai-in-psychiatry-detecting-mental-illness-with-artificial-intelligence/95028/

Rayome, A. D. (2020, April 21). This robot could make COVID-19 testing faster and safer. Retrieved July 24, 2020, from https://www.cnet.com/news/this-robot-could-make-covid-19-testing-faster-and-safer/

Richardson, J. (2017, May 03). Work gives our lives meaning. What will we do when robots have taken our jobs? Retrieved July 31, 2020, from https://qz.com/973149/work-gives-our-lives-meaning-what-will-we-do-when-robots-have-taken-our-jobs/

Riley, T. (2018, May 09). Can Increased Automation Negatively Affect People's Physical and Mental Health? Retrieved July 31, 2020, from https://psmag.com/social-justice/automation-increases-anxiety

Robotopia. (2020, March 18). What America can learn from China's use of robots and telemedicine to combat the coronavirus. Retrieved from https://www.cnbc.com/2020/03/18/how-china-is-using-robots-and-telemedicine-to-combat-the-coronavirus.html

Robots also adapt to the world of COVID-19. (2020, June 06). Retrieved from https://atalayar.com/en/content/robots-also-adapt-world-covid-19

Romero, M. E. (2020, April 8). Tommy the robot nurse helps Italian doctors care for COVID-19 patients. The World from PRX. https://www.pri.org/stories/2020-04-08/tommy-robot-nurse-helps-italian-doctors-care-covid-19-patients

Rutgers University. (2020, February 6). New robot does superior job sampling blood: First clinical trial of an automated blood drawing and testing device. ScienceDaily. https://www.sciencedaily.com/releases/2020/02/200206132343.htm

Sanders, R. (2020). "UC Berkeley scientists spin up a robotic COVID-19 testing lab". Berkeley News. Retrieved from https://news.berkeley.edu/2020/03/30/uc-berkeley-scientists-spin-up-a-robotic-covid-19-testing-lab/

Sascha, Keutel. (2020). Robots help fight hospital infections. Healthcare in Europe. Retrieved from https://healthcare-in-europe.com/en/news/robots-help-fight-hospital-infec
Tions.html

Shafiev, F. (2020, April 28). The Architecture of International Relations After COVID-19: A Return to the 'New Normal'. Retrieved August 4, 2020, from https://valdaiclub.com/a/highlights/the-architecture-of-international-relations-after-/

Shortage of personal protective equipment endangering health workers worldwide. (2020, March 3). Retrieved from https://www.who.int/news-room/detail/03-03-2020-shortage-of-personal-protective-equipment-endangering-health-workers-worldwide

Smith, L. (2020, March 27). How robots helped protect doctors from coronavirus. Fast Company. https://www.fastcompany.com/90476758/how-robots-helped-protect-doctors-from-coronavirus

Solis, B. (2014, September 09). COVID-19 accelerates enterprise use of automation in digital transformation. Retrieved July 24, 2020, from https://paperzz.com/doc/1146164/breast-cancer-research-and-treatment

Speer, J. (2020, April 28). Intelligent Robotics: What to Expect in the Post-COVID-19 Era. Retrieved August 4, 2020, from https://www.mhlnews.com/technology-automation/article/21129896/intelligent-robotics-what-to-expect-in-the-postcovid19-era

Swenson, C. A., & Quinn, T. J. (2004, June 17). Thermometry. Retrieved July 17, 2020, from https://www.sciencedirect.com/topics/chemistry/thermometry

Szalavitz, M. (2010, March 01). Touching Empathy. Retrieved July 31, 2020, fromhttps://www.psychologytoday.com/ca/blog/born-love/201003/touching-empathy

Tiernan, K., & Ahmadinejad, S. (2020). ROBOTIC PROCESS AUTOMATION DURING COVID-19 [PDF]. Zaventem: Binder Dijker Otte.

Tirumalaraju, D. (2019, December 24). Novartis partners with Amazon for digital boost to manufacturing. Pharmaceutical Technology. https://www.pharmaceutical-technology.com/news/novartis-manufacturing-aws/

Transcat (2018, January 1). Questions and Answers about Infrared

Thermometry. Retrieved
July 17, 2020, from https://www.transcat.com/calibration-resources/application-notes/infrared-thermometry

University of California San Diego. (2020). How robots can help combat COVID-19. Retrieved from How robots can help combat COVID-19.

Vincent, J. (2018, January 23). Artificial intelligence is going to supercharge surveillance. Retrieved July 24, 2020, from https://www.theverge.com/2018/1/23/16907238/artificial-intelligence-surveillance-cameras-security

Wang, S., Zha, Y., Li, W., Wu, Q., Li, X., Niu, M., . . . Tian, J. (2020). A Fully Automatic Deep Learning System for COVID-19 Diagnostic and Prognostic Analysis. European Respiratory Journal, 2000775. doi:10.1183/13993003.00775-2020

Walker, C. S. (2020, April 29). How personal contact will change post-Covid-19. Retrieved August 4, 2020, from https://www.bbc.com/future/article/20200429-will-personal-contact-change-due-to-coronavirus

Wiggers, K. (2020, March 24). Miso Robotics deploys AI screening devices to detect signs of fever at restaurants. Retrieved from https://venturebeat.com/2020/03/24/miso-robotics-deploys-ai-screening-devices-to-detect-coronavirus-covid-19-fever/

Wilkerson, B. (2019, May 26). Opinion: Artificial intelligence could have damaging effects on our mental health. Retrieved July 31, 2020, from https://www.theglobeandmail.com/business/commentary/article-artificial-intelligence-could-have-damaging-effects-on-our-mental/With the outbreak and long-term Impact of COVID-19, Global Smart Home Automation Market to reach USD 149.8 billion by 2027. (2020, June 19). Retrieved from https://www.marketwatch.com/press-release/with-the-outbreak-and-long-term-impact-of-covid-19-global-smart-home-automation-market-to-reach-usd-1498-billion-by-2027-2020-06-19?tesla=y

Witte, K. (2019, July 30). Effects of automation on unemployment and mental health: UBI. Retrieved July 31, 2020, from https://doc-research.org/2019/07/automation-unemployment-mental-health-ubi/

Worldometer. (2020). COVID-19 Coronavirus Pandemic. Retrieved from https://www.worldometers.info/coronavirus/

Writer, S. C. (2020, February 14). Report Reveals Worst State for Healthcare Data Breaches in 2019. Retrieved from https://www.infosecurity-magazine.com/news/report-healthcare-data-breaches-in/#:~:text=The report found that 510,2018 to 41,335,889 in 2019.

Yum, S. (2020). Social Network Analysis for Coronavirus (COVID☒19) in the United States. Social Science Quarterly. doi:10.1111/ssqu.12808